T0243855

JOHN HARRISON

and the Quest for Longitude

2nd Edition

JONATHAN BETTS

First published in 1993 by the National Maritime Museum,
Park Row, Greenwich, London, SE10 9NF.

This expanded and revised edition published in 2023.

Produced with support from the Antiquarian Horological Society (www.ahsoc.org)

ISBN: 9781906367992

At the heart of the UNESCO World Heritage Site of
Maritime Greenwich are the four world-class attractions of
Royal Museums Greenwich – the National Maritime Museum,
the Royal Observatory, the Queen's House and *Cutty Sark*.

www.rmg.co.uk

A CIP catalogue record for the book is available from the British Library.

Designed by Megan Sheer (sheerdesignandtypesetting.com)

Technical illustrations by Lee Yuen-Rapati

Indexing by Jonathan Eyers

Printed and bound in the UK on FSC certified paper by Gomer Press

CONTENTS

ACKNOWLEDGEMENTS

Principal among our little band of Harrison researchers was the late Martin Burgess, whose never-flagging enthusiasm, along with his own Harrison-inspired clocks, were an inspiration to us all. In researching Harrison's life, and particularly his wooden pendulum clocks, Andrew King has contributed enormously to our knowledge. Andrew has also read through the text of this publication and made many very helpful comments and corrections, for which I am exceedingly grateful. Thanks are also due to the Antiquarian Horological Society for their support of this edition and the photo studio and publishing teams of Royal Museums Greenwich for seeing this book through to completion.

This edition is dedicated to the memory of Martin Burgess (1931–2022).

PREFACE

This book tells the story of John Harrison (1693–1776), perhaps history's most famous watch and clockmaker. Since Dava Sobel's bestseller *Longitude* was first published in 1995, Harrison has become much better known, but one thing has always been needed: a simple and accessible description of Harrison's creations, set alongside the story of his life. Harrison's marine timekeepers and pendulum clocks helped solve one of the greatest scientific problems of the age – how to find longitude at sea – and it is in these extraordinary machines that his real achievement lies.

The story of how the timekeepers were made is also a great tale of triumph over adversity. Although in the early years of his research and development work in London Harrison was given considerable encouragement by a number of important figures in the scientific community, in later years there was significant prejudice about the viability of the timekeeper solution and he faced many practical and technical difficulties along the way. In spite of this, Harrison's work produced one of the major technological achievements of the eighteenth century. Indeed, his invention, a truly accurate portable timekeeper, was not just the first practical *marine chronometer* but also the foundation of all subsequent precision watches.

A superficial look at his clocks suggests incredibly complex mechanisms, impossible for any but professionals to understand. Many visitors to the Royal Observatory tell us that in fact they admire them primarily as works of art: they are such strikingly beautiful things to see running that they are mainly appreciated as kinetic sculpture. They may be complex to look at, but their function is actually very simple and easy to understand: their sole purpose was to tell the time as accurately as possible. They do not strike the hours, chime the quarters or play music, nor do they have an alarm to wake you up in the morning, as many contemporary clocks and watches did. They do not have complicated calendar indications, automata mechanisms or astronomical indications to amuse, impress and educate.

They do have lots of parts, it is true; and it is also true that those parts were designed and created by Harrison in ways that were unique to him. Being self-taught, and working almost entirely alone, his scientific and horological thinking was deeply unconventional, and this is one of the most interesting and challenging aspects of studying his mechanisms and writings. Harrison was also a pioneer in his field; it could be said that his was one of the first government-sponsored research and development projects in England so, by definition, many of his mechanisms and concepts are unique.

It is said that imitation is the sincerest form of flattery, and since the 1970s there have been a number of replicas made of Harrison's timekeepers and pendulum clocks. Although cared for by the National Maritime Museum, the marine timekeepers H1 to H4 were, until 2017, the property of the Government, and the official policy of the Ministry of Defence was always to discourage replicas on the grounds that poor copies would demean the originals. In recent years, however, some very fine replicas have been made by independent craftsmen, even without access to the originals. Currently three full-size and one half-size copy of H1 are known, a copy of H2, one of H3 and one of H4 were completed in recent years and copies of the Royal Astronomical Society regulator are underway. There is no doubt that a great deal can be learned from such projects and, particularly with the replica of H4, it will be fascinating to hear how well they perform.

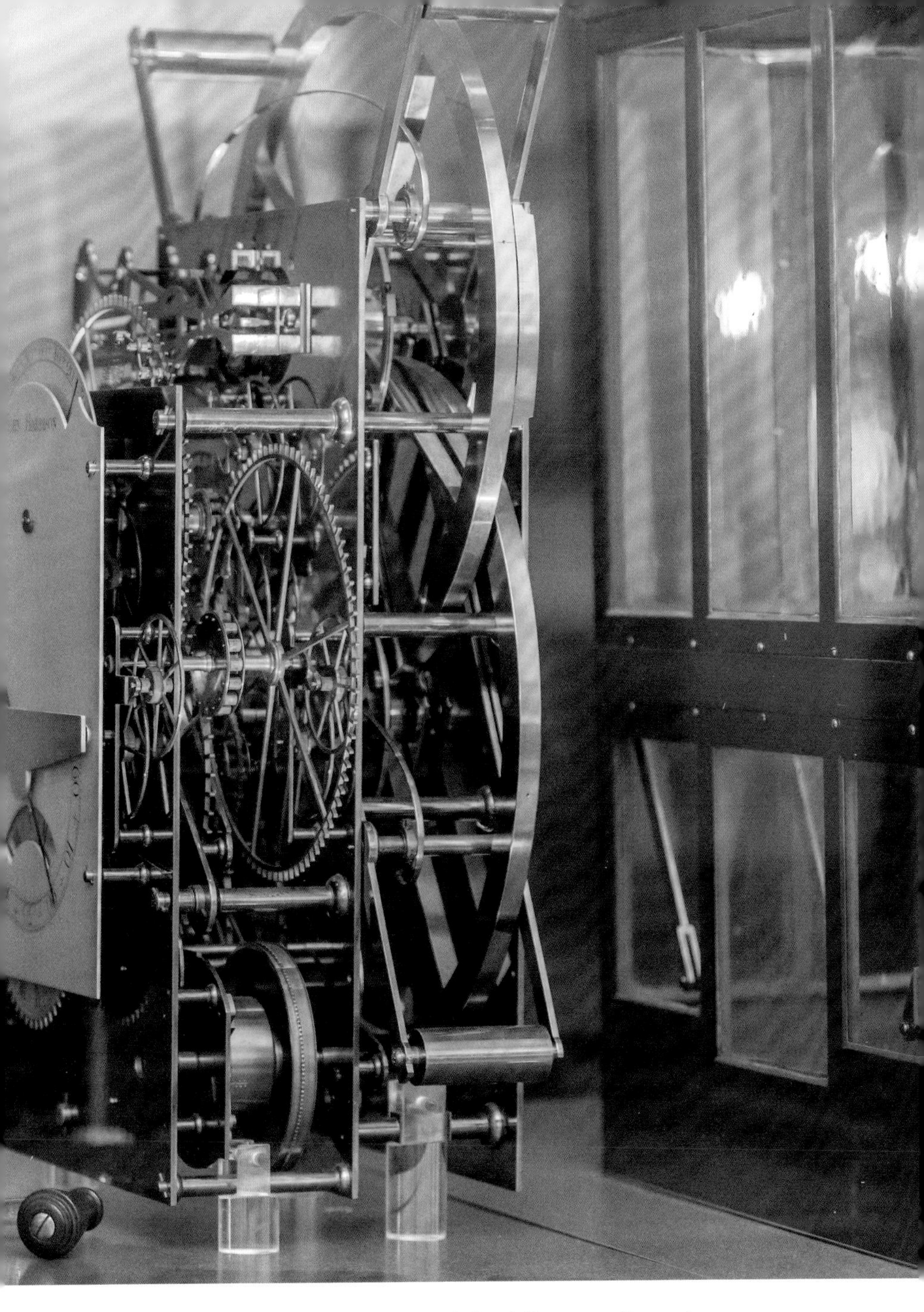

H3 on display in the 'Time & Longitude' gallery at the Royal Observatory, Greenwich.

A BRIEF INTRODUCTION TO CLOCKWORK

John Harrison was a pioneer in the field of precision clockmaking, but he was by no means the first to make mechanical clocks. At the time he was making his first timepiece, mechanical clocks had been around for over 400 years; there is evidence that clockwork was being made in thirteenth-century Europe. The first examples appear to have been large mechanisms, made in monasteries, as a means of automatically sounding bells for prayer (the term clock derives from the Latin for bell, 'clocca'); made without dials, they sounded the time rather than showing it.

The mechanism of H1, Harrison's first marine timekeeper.

At the beginning of the fourteenth century, clocks with dials to show the time began to appear in communities across central Europe, as a convenient alternative to the sundial. They were especially useful in cloudy weather but particularly for timekeeping during the night. Of relatively crude construction, made almost entirely of iron and wood, and powered by a weight, these early machines were very poor timekeepers. At the end of a day they might be out by as much as half an hour, and, during the first 350 years that clockwork was being made, a sundial was far more trustworthy than a clock. At the time, this did not matter much, as there was usually only one clock in any given community and, right or wrong, at least everyone agreed what time it was!

In the late fourteenth century, smaller versions of the large mechanical clocks appeared, made for domestic use, some as alarm clocks only running overnight, and some for timekeeping and striking the hours all day. Then, around 1450 or soon after, there appeared the concept of a coiled spring (instead of a weight hanging on a cord) for driving a clock. For the first time, clocks didn't need to be fixed high up on a wall and could be carried about.

By 1500, during the early Renaissance, cities in southern Germany such as Nuremburg and Augsburg had established themselves as European centres for fine metalwork, and it was here that clockwork began to be made so small that it could be carried on the person: watches had arrived. Although very expensive to produce, these fine articles of jewellery still kept time no better than before. They were chiefly bought by the wealthy as tokens of prosperity and status, and particularly as symbols of learning and temperance. All this was to change in the following century, as in Europe an increasingly rational, free-thinking view of science and theoretical and practical experiments would completely revolutionise horological technology.

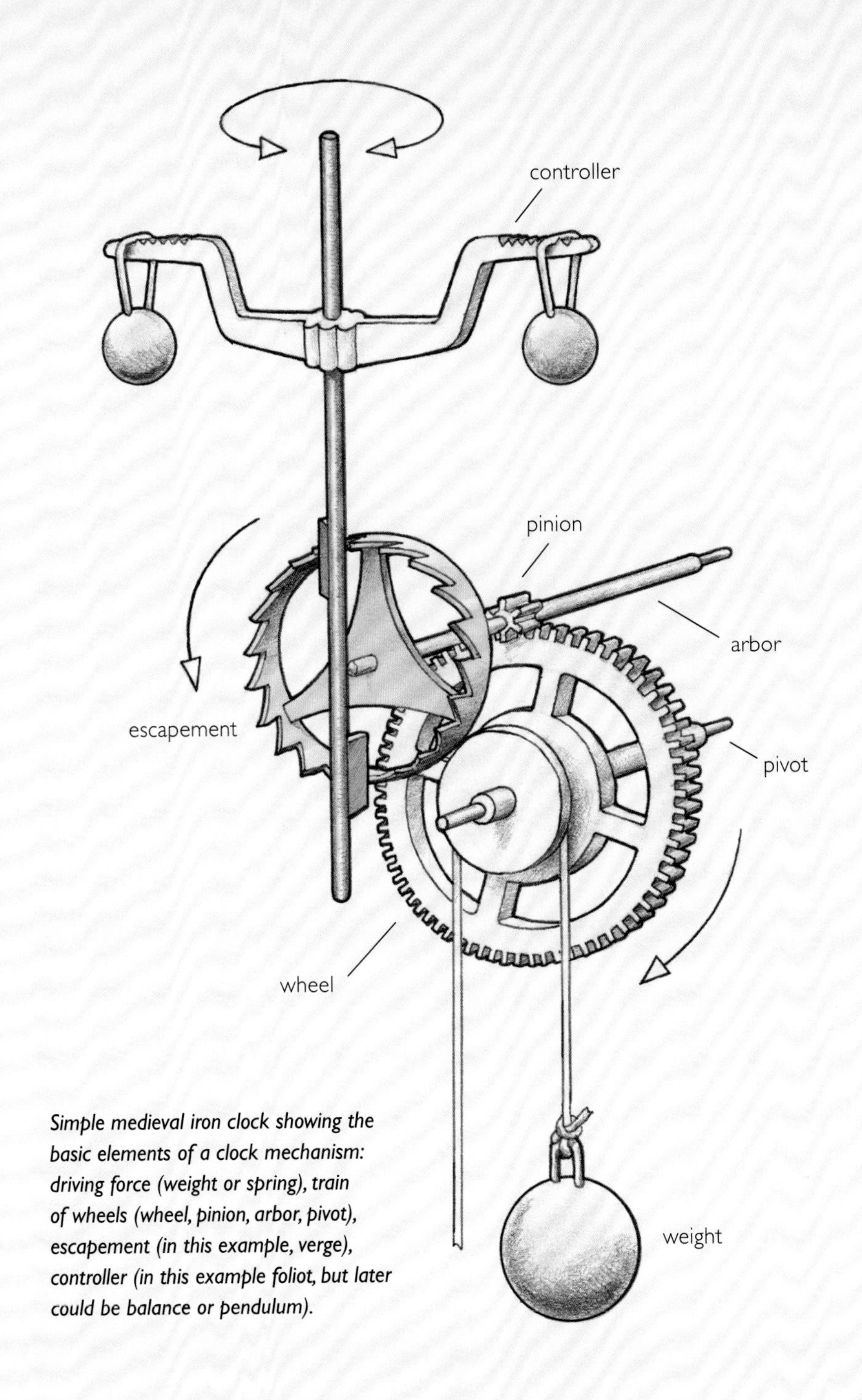

Simple medieval iron clock showing the basic elements of a clock mechanism: driving force (weight or spring), train of wheels (wheel, pinion, arbor, pivot), escapement (in this example, verge), controller (in this example foliot, but later could be balance or pendulum).

These diagrams show the main parts of a clock. The mechanism is generally known as the ***movement***.

The ***frame*** holds all the working parts of the timekeeper together; in most clocks and watches from the sixteenth century onwards, this is made in brass, consisting of a pair of plates, held apart by pillars.

At the bottom end of the movement is the power source which is the element that drives the clock, usually a ***weight*** suspended on a cord or a coiled ***mainspring***.

The ***wheels*** in the movement, which feed the energy to the timekeeping element and supply the correct motion to the hands and dial of the clock, are known as the ***train***. Most of the wheels consist of an axle, the ***arbor***, with a large toothed wheel and a smaller toothed ***pinion*** mounted on it. The arbors of the train have at each end a little extension called a ***pivot*** which, in conventional clocks, runs in a ***pivot-hole*** in the frame of the clock.

Finally there are the all-important duo, the ***escapement*** and ***controller***, grand-sounding names for the most important elements that make up the 'beating heart' of the movement, the part that actually keeps the time. The controller in early clocks was a simple, horizontal bar called a ***foliot***, which swung to and fro at the speed determined by the force from the wheels of the clock. Later the foliot was replaced with a simple wheel called a ***balance***, and later still, the ***pendulum*** was introduced as a much more accurate form of controller.
The escapement is the part that feeds in the energy to keep the controller swinging, whether it is pendulum or balance. Common longcase clocks in Harrison's day used an ***anchor escapement***, so-called because the ***pallets***, the parts that receive the push from the wheels of the clock, look like an inverted anchor.

Verge escapement with a balance, as fitted to clocks and watches.

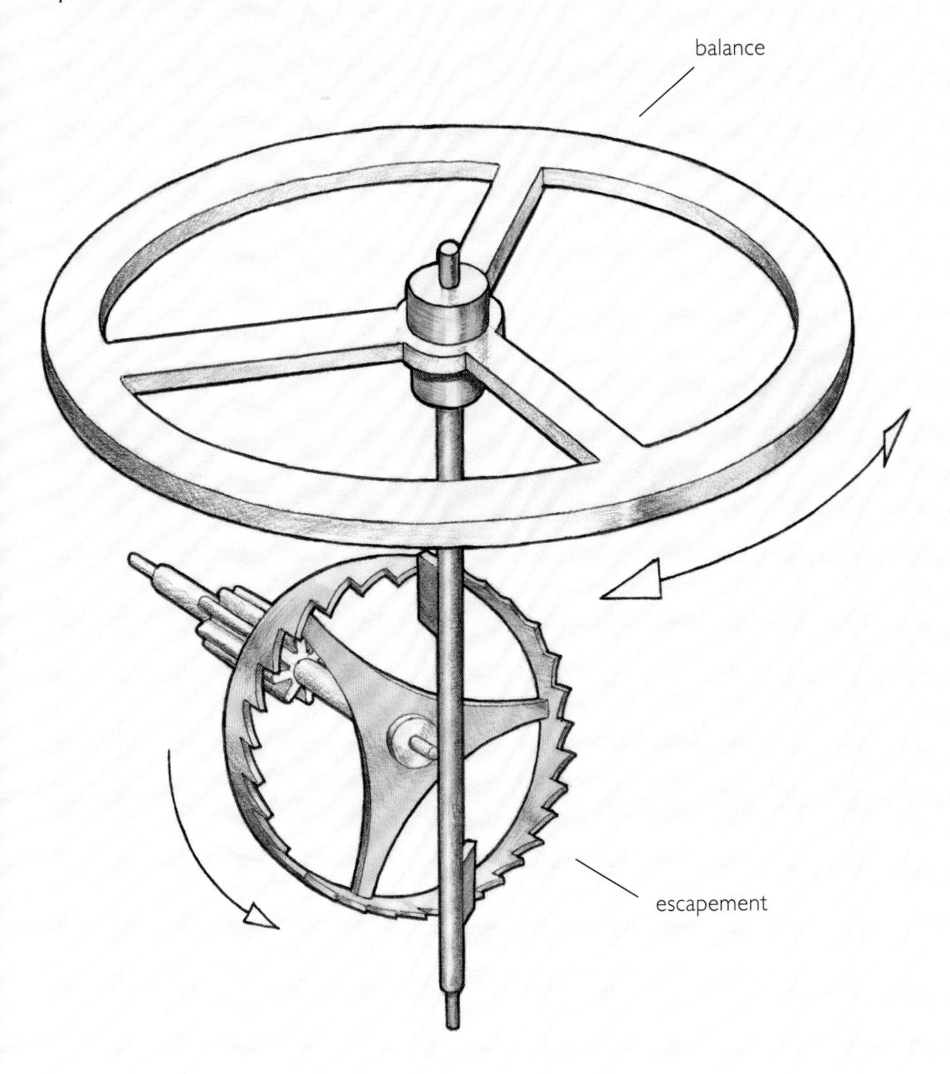

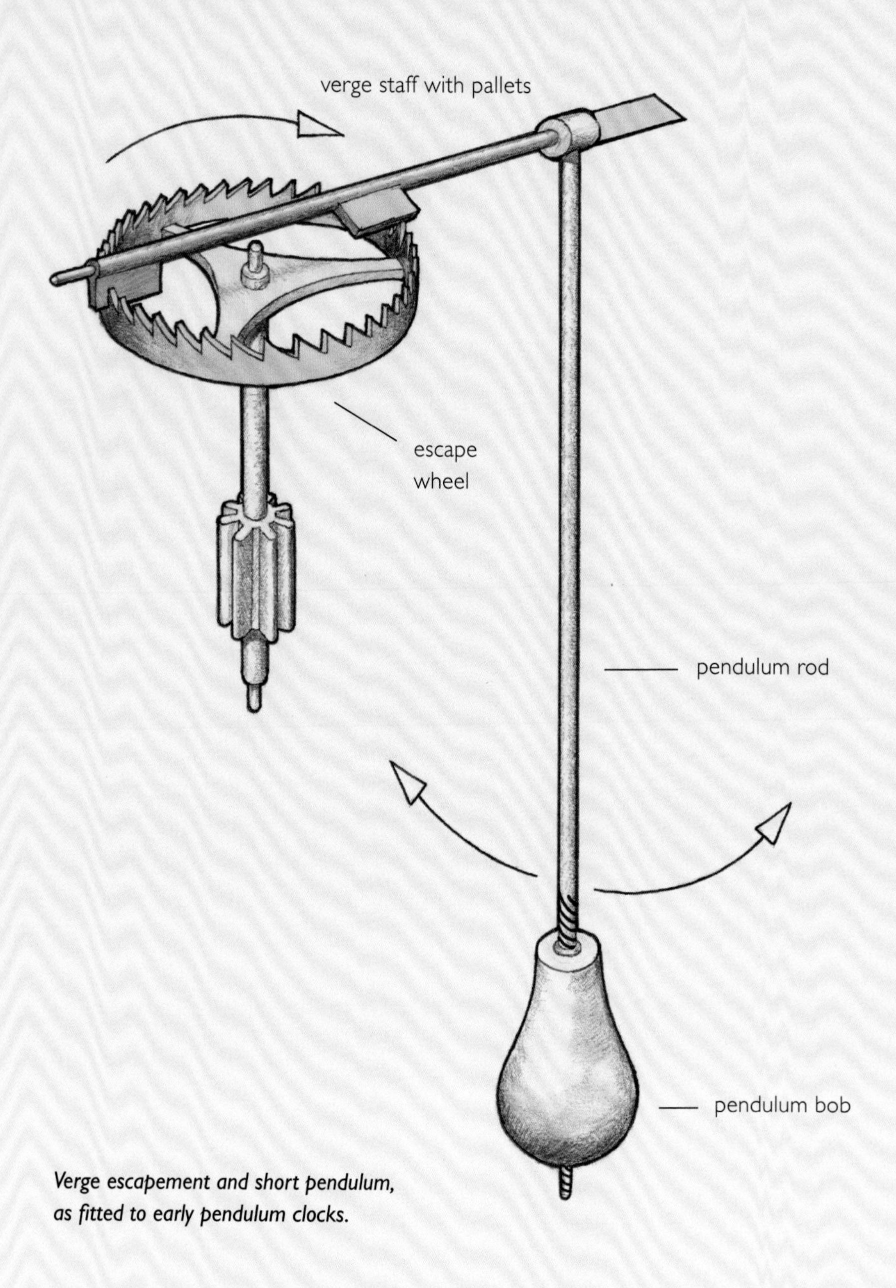

Verge escapement and short pendulum, as fitted to early pendulum clocks.

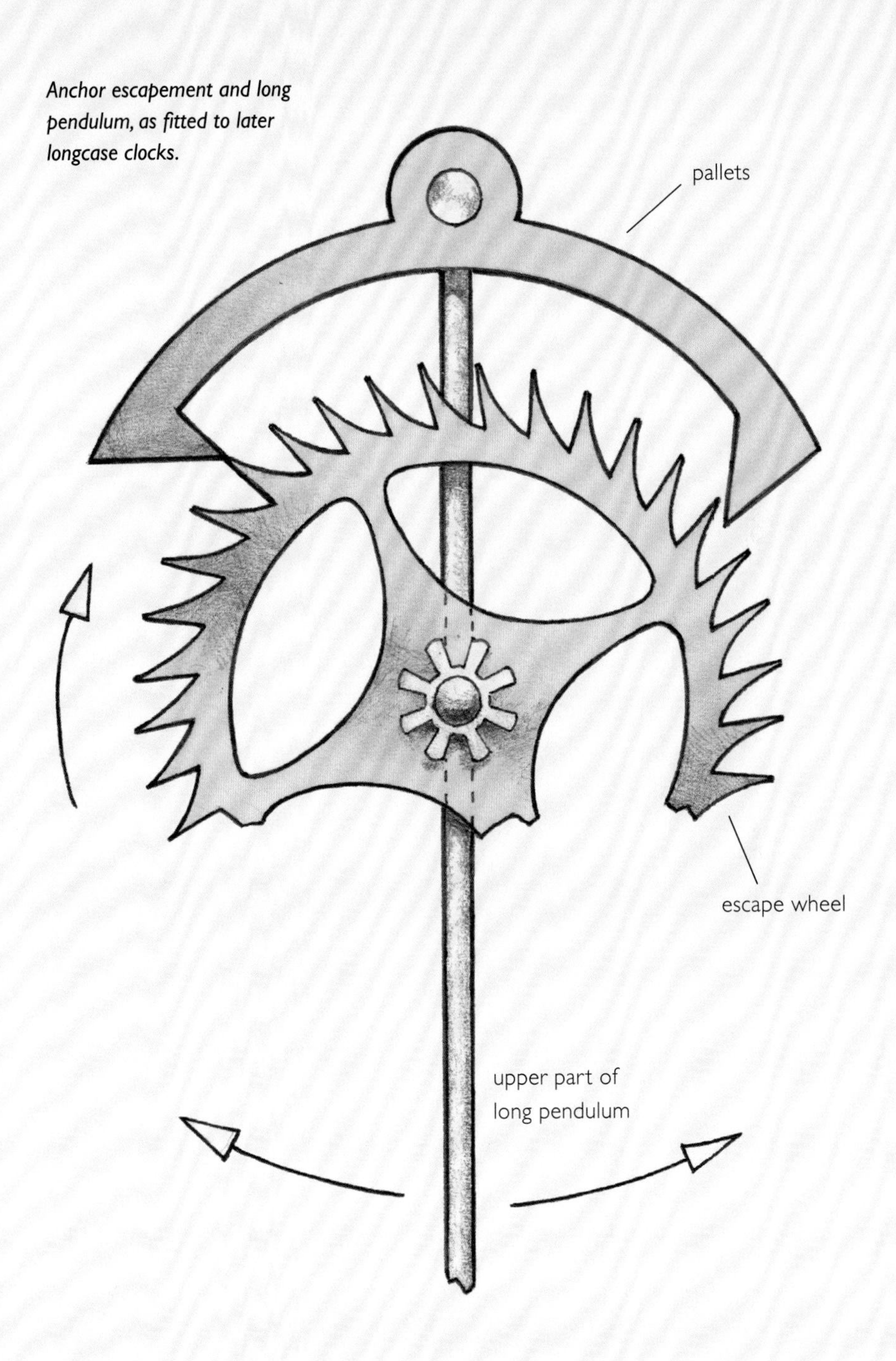

Anchor escapement and long pendulum, as fitted to later longcase clocks.

The spring in a spring-driven clock delivers a force that is much greater when wound up than when wound down and so, from the earliest period of clockwork, spring-driven movements were almost always fitted with a means to equalise the force, usually a device known as a ***fusee***. This is a conical pulley (a bit like a tiny fairground 'helter-skelter' with a cord wrapped round it), which ensures a uniform driving force to the wheels of the clock. The mainspring is fitted into a drum (the barrel), around which is a cord that then wraps around the conical fusee. The fusee has the first wheel in the train attached to its base. When the clock is fully wound, and the spring is pulling at its greatest force, the cord pulls on the smallest radius of the fusee; but as the clock runs, and the fusee slowly rotates, so the cord pulls on a greater and greater radius, giving an increasing leverage, at the same time as the mainspring's force diminishes. If matched correctly, the fusee can almost completely equalise the driving force delivered to the wheels.

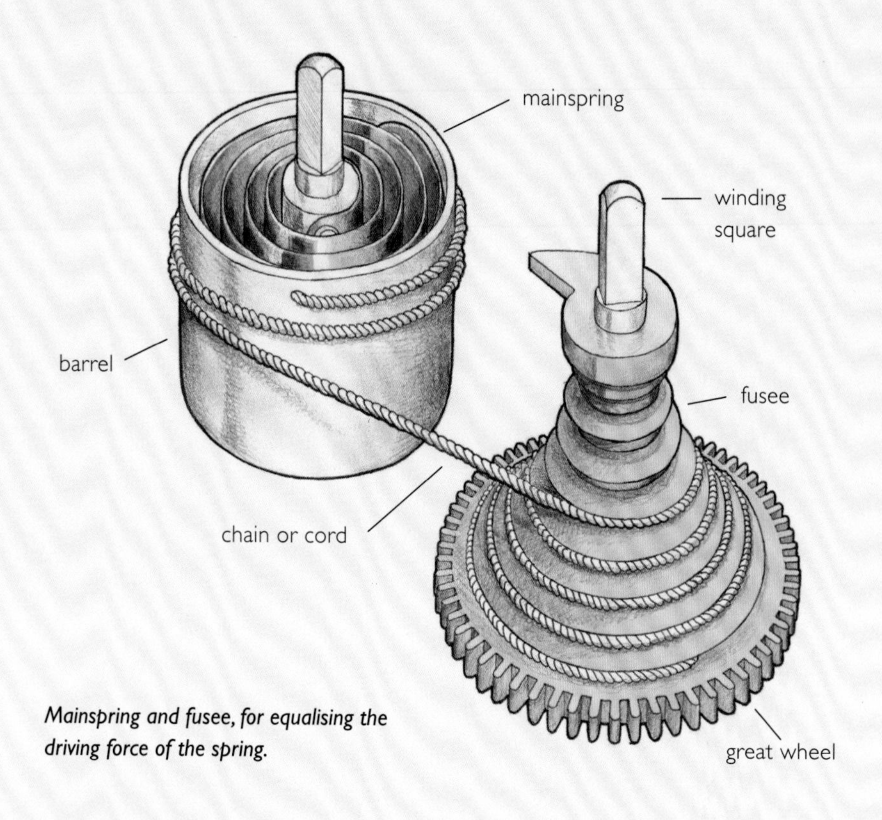

Mainspring and fusee, for equalising the driving force of the spring.

A SCIENTIFIC GOLDEN AGE

Throughout the second half of the seventeenth century, England too enjoyed this scientific 'Golden Age'. Under King Charles II's patronage, the Royal Society was founded in London in 1660 and, with his funding, the Royal Observatory was built at Greenwich in 1675. The work of scientists and mathematicians such as Isaac Newton (1642–1727), Robert Boyle (1627–91) and Robert Hooke (1635–1703) in England reflected a period of intense scientific advancement across northern Europe.

Inspired by the celebrated Italian astronomer Galileo Galilei's (1564–1642) brilliant realisation that a swinging pendulum would make a highly accurate controller for a timekeeper, the great Dutch scientist Christiaan Huygens (1629–95) designed the first practical pendulum clock in 1656. While the timekeeping of earlier clocks had been in error by about 20 to 30 minutes a day, the new pendulum-controlled clocks were capable of a variation of less than a minute a day: the introduction of the pendulum to clockwork was one of the greatest improvements in the history of timekeeping. The reason the pendulum is such a good timekeeper is that, unlike the controllers in earlier clocks (the foliot or simple balance wheel), the pendulum has a natural 'restoring force' – gravity – which ensures it swings at a very regular rate.

Clockmakers in Holland and France began producing these technological wonders, but it was London clockmakers, most notably Ahasuerus Fromanteel (1607–93), who had sent his son John Fromanteel (*c.*1638–*c.*1690) to Holland to learn of Huygens's invention, who exploited the new design to its maximum potential.

In 1675 a similar invention occurred in watchwork – the application of the *balance spring*. From this time watches too became more accurate. With the work of these pioneers and the next generation, including the great Thomas Tompion (1639–1713), 'the father of English watchmaking', there began a horological 'Golden Age' to complement the scientific one in England.

By the end of the seventeenth century, London was universally recognised as the world's most important centre for the manufacture of clocks and watches. Virtually all the important design improvements that shaped the modern mechanical clock were incorporated in English clocks by 1700. Only one real challenge remained to be met in improving clockwork, and it was no wonder the world looked to England for a solution.

THE BALANCE SPRING

Soon after the invention of the pendulum for clockwork, a similar improvement was introduced for watches. Up until then, the timekeeping element in watches had been a simple small balance wheel, a tiny version of the kind found in clockwork, swinging to and fro, only controlled by the force from the mainspring of the watch. Now, just as the new pendulum clocks had gravity to control their oscillations more consistently, watches had a little coiled spiral spring applied to the balance, giving its swings a natural frequency too. The principle of applying a spring to the watch balance was conceived by the English scientist Robert Hooke in the 1660s, who described the effect as like applying an 'artificial gravity' to the oscillator. But it was only the spiral form of the spring that was the first really workable design and this was created by Christiaan Huygens around 1675. Now watches could keep time to about a minute a day, a huge improvement, reminiscent of that in pendulum clocks, and one which led to them sometimes being referred to as 'pendulum watches'.

Balance fitted with a balance spring, in use from 1675.

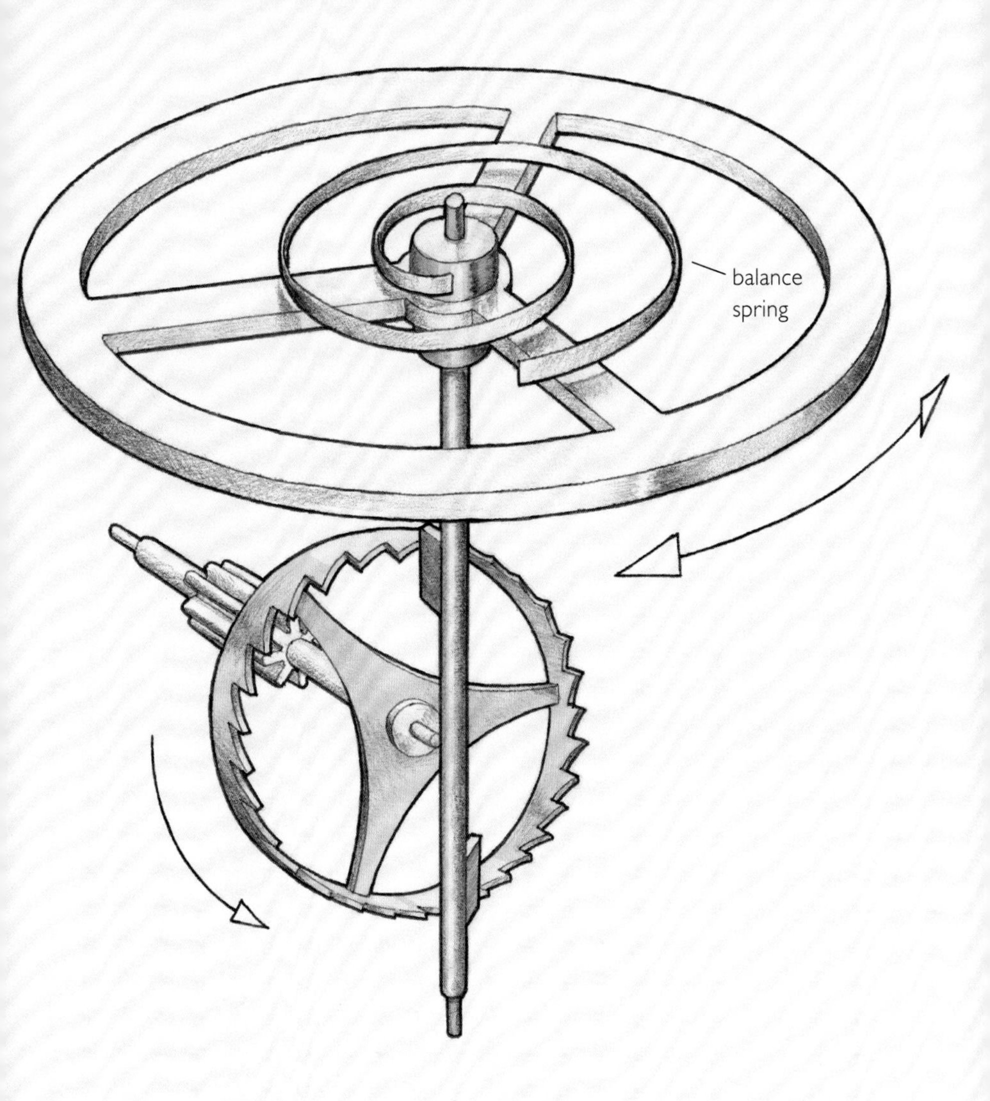

The Wreck of the 'Amsterdam',
attributed to Cornelis Claesz van Wieringen, c.1630.

THE LONGITUDE PROBLEM

From the end of the fifteenth century, merchants, explorers and adventurers took to the open seas in unprecedented numbers. These journeys were hazardous not only because of the inherent dangers of the sea but also because, once out of sight of land, sailors had no accurate means of knowing their exact position.

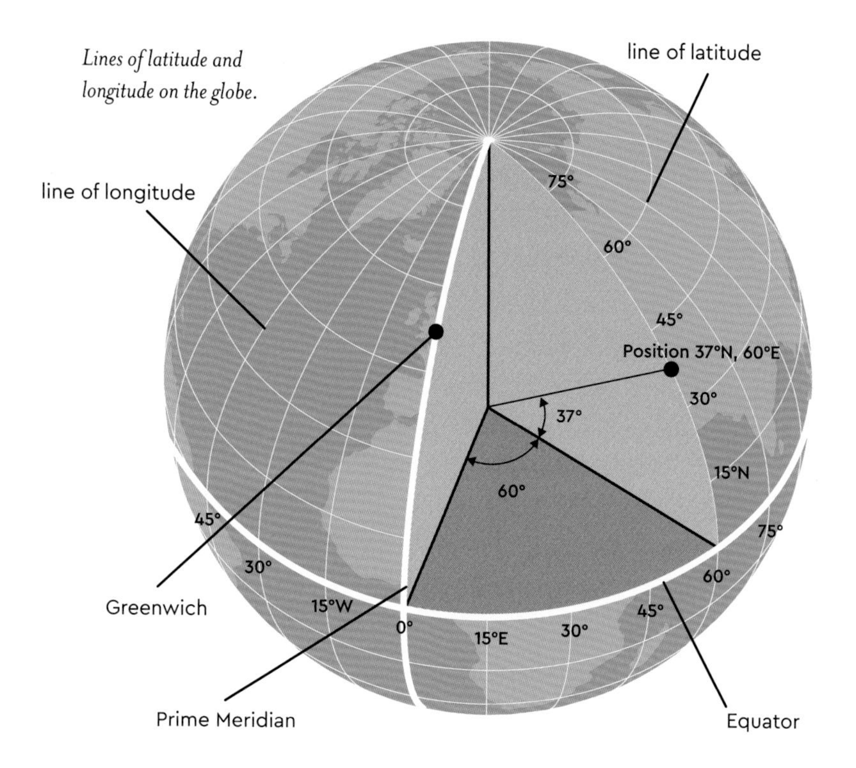

Lines of latitude and longitude on the globe.

One's position on Earth is generally defined by two co-ordinates: *latitude*, one's distance north or south of the Equator, and *longitude*, one's distance east or west of an agreed place, such as one's home port – in other words how far around the world one is from home. Latitude was easy to find, with a little calculation, by observation of the Sun at midday or (in the northern hemisphere) by the Pole Star at night. Longitude, however, had always been a problem.

Because Earth revolves on its axis, the Sun appears to traverse the sky from east to west. One complete revolution of Earth takes one day (24 hours) and we are familiar with the fact that today the world is divided into 24 *time zones*. In the same way, the circle of the globe can be divided into 360° of longitude so, as the 24 hours of time are the same thing as 360° of longitude, each hour is equivalent to 15° of longitude.

Represented by lines known as *meridians* – running from pole to pole, dividing Earth into these equal segments – longitude difference is thus the same thing as *time difference* between places. For example, when one calls New York from Greenwich and it happens to be 12 noon in Greenwich, we know that at that same moment it will be only 7 a.m. in New York. Therefore, the time difference of 5 hours behind means that New York is 75° of longitude west from Greenwich.

So, to find your longitude from home when out at sea, you only need to know two things: your local time (which is relatively easy to find, using the Sun or the stars) and the time at your home port at that same moment. The difference between the two times then provided you with the longitude from home.

The real problem was how to discover what time it was at the home port. The obvious answer would be to take a portable clock with you which was set to 'home time' before you left. Such a clock must never be allowed to stop, and would have to be both very accurate and unaffected by the movements of the ship and temperature changes. In 1700 almost no one believed such a clock could be made.

Having designed the highly successful and practical pendulum clock, Christiaan Huygens made considerable efforts in the seventeenth century to develop a seagoing clock, as did his English contemporary Robert Hooke and a number of other hopefuls. But any marine clock design that relies on gravity for its timekeeping is bound to fail, owing to the widely varying forces on board a moving ship, and pendulum clocks could not possibly function reliably under those conditions. Even the great scientist Sir Isaac Newton considered the mechanical clock solution most unlikely to succeed.

One alternative to the 'impossible clock' was to use the movement of the Moon in the sky as a kind of celestial 'clock' to provide 'home time'. Since its passage past the stars was very predictable, the Moon could be used like the clock's 'hand', with the stars as the clock's 'dial'. By consulting tables listing the position of the Moon at different 'home times', one could calculate one's longitude when at sea. Unfortunately, though, this system – called the *lunar-distance* method – was complex and time-consuming. It could take four hours to find longitude this way, besides which, for 20 per cent of the month, the Moon could not easily be seen. Nevertheless, many people initially believed that the lunar-distance method would be the only answer to the longitude problem. Indeed, the Royal Observatory was founded at Greenwich in 1675 for the express navigational purpose of making charts of the skies and to record the position of the Moon throughout the year, 'so as to find the so-much-desired longitude of places for perfecting the art of navigation', as the Observatory's charter states.

The first Royal Observatory building, designed by Sir Christopher Wren in 1675 and known today as Flamsteed House. Engraving by Francis Place, c.*1676.*

THE GREAT LONGITUDE REWARDS

Meanwhile, seafaring nations continued to sail the oceans in spite of the longitude problem, and many lives and large quantities of cargo were still being lost every year. As early as 1567 the Spanish Crown had offered monetary reward for a solution to this seemingly intractable problem and, in the following century, further prizes were offered by the governments of Portugal, Venice and Holland. King Louis IV of France himself had tentatively agreed to reward one particular inventor who proposed a solution, but it was eventually deemed to be unsatisfactory, and the whole question of determining longitude at sea was considered an impossible task.

The year 1707 witnessed the calamitous loss of the British flagship *Association* alongisde three other warships, wrecked on the Isles of Scilly, just off the south-west coast of England. The Admiral of the Fleet, the splendidly named Sir Cloudesley Shovell (1650–1707), lost his life, along with nearly 2,000 of his men, a disaster as shocking in its day as the loss of the *Titanic* in 1912. Although the principle cause of this terrible accident was not an inability to determine longitude (the fleet was in fact much further north than it had estimated), it served as a dramatic and awful reminder of the need for safer navigation at sea, and undoubtedly partly inspired the developments that followed.

In July 1713, the noted mathematicians William Whiston (1667–1752) and Humphrey Ditton (1675–1715) published a letter in the London-based journal *The Guardian* stating that they

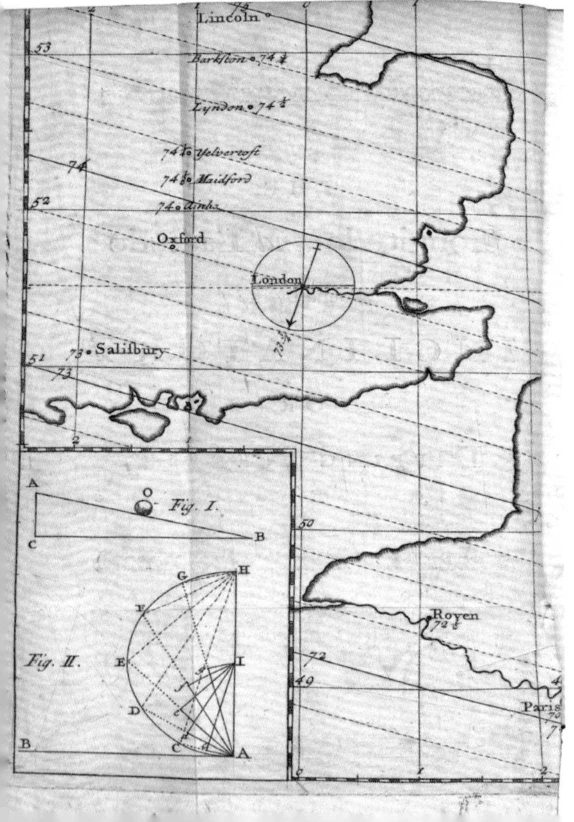

Above: *The Admiral of the Fleet, Sir Cloudesley Shovell, and nearly 2,000 men lost their lives when four navy ships were wrecked off the Isles of Scilly on 22 October 1707. Artist unknown, c.1710.*

Left: *Another longitude proposal of William Whiston, published in 1719, included using the variation of the compass as a solution. Across the globe, magnetic North (by a compass) differs from the true North (by the pole star) by different amounts. Whiston proposed that navigators could use this difference to find their position on Earth.*

had a solution to the longitude problem and would disclose it if a reward were granted, subject to the inspection and approval of Sir Isaac Newton. The method itself, which proposed 'fixed' barges, 'moored' at regular intervals across the Atlantic Ocean and from which 'rocket' time signals would be fired, was not practical, to say the least! Nevertheless, Whiston and Ditton were good publicists, as well as friends of Newton's, and the proposal gave the issue of longitude a much higher profile. Pressure was now mounting from influential merchants, ships' captains and commanders alike for a solution to be found, and the following year the Government responded by passing an Act of Parliament offering large rewards for a solution to the longitude problem (see p. 66).

£10,000 was on offer for a solution that could determine longitude to within one degree, and £15,000 for one within two-thirds of a degree.

The greatest reward was £20,000 (equivalent to more than £2,000,000 today) and to qualify for this, the solution (any method would be considered as long as it was practical) had to provide longitude to within half a degree. To test the method it would be tried on a ship, sailing 'over the Ocean, from Great Britain to any such Port in the West Indies, as those Commissioners, or the major part of them, shall Choose [...] without losing their Longitude beyond the limits before mentioned'. The final requirement was that the method should prove to be 'tried and found Practicable and Useful at Sea'. Earlier discussions had repeatedly focused on theoretical solutions to the longitude problem, and this statement in the Act simply meant that the successful method must not be just a theory, but of proven use by real mariners on a real ship at sea. Inasmuch as that point was clear, the Act – the last to be given royal assent by Queen Anne – was simple and unambiguous in its terms.

THE IMPOSSIBLE QUESTION

One unfortunate result of the Act was that the offer of such large sums of money attracted the attention of all manner of cranks and charlatans. The official body set up to judge the submissions and administer the rewards, later titled The Board of Longitude and made up of senior politicians, high-ranking naval personnel and 'Oxbridge' professors, received many weird and wonderful proposals. Whiston and Ditton's ideas must have seemed perfectly sane by comparison! The whole longitude question was soon so famous for its difficulty that, like squaring the circle and inventing a perpetual-motion machine, it became a kind of catchphrase for the pursuits of fools and lunatics. Many simply believed the problem could not be solved and, over the years, commentators often descended to ridicule and invoked heavy sarcasm when referring to the subject.

Perhaps the most famous example of this was the tongue-in-cheek publication by an anonymous author in 1688 – decades before the Act of Parliament – of a proposal using the 'Powder of Sympathy'. This substance was the brainchild of Sir Kenelm Digby (1603–65), who had claimed that patients who had suffered wounds caused by knife or sword were cured simply by placing a bandage from the wound in the mysterious powder. Digby stated that when this was done, the patient immediately felt the effect on the wound, even though they were sometimes a long distance away. The anonymous author of the spoof longitude proposal suggested that every ship be provided with a dog, suitably 'wounded', and, at appropriate times

The 'longitude lunatics' in Bethlem Hospital, from The Rake's Progress *by William Hogarth, 1735. The figure by the wall attempts to solve the longitude problem by drawing, while another looks at the stars through a paper 'telescope'.*

back at home, a bandage or knife associated with that wound be plunged into the powder of sympathy, causing the dog to jump and providing a kind of canine time-signal across the miles! It was, of course, a joke, but, to many, the idea that a clock might succeed where dogs had failed was equally ridiculous.

The principal purpose of the Act of Parliament was thus to discover whether any solution was possible at all. If it could be proven that a method had the potential to succeed, as shown by the specified successful trial voyage to the West Indies, then even the full reward of £20,000 would be money well spent. For most people such a sum meant unimaginable riches, but in terms of naval budgets it represented less than half the cost of a second-rate ship of the line and just one ship saved from loss would amply repay the cost of the reward.

MARINE TIMEKEEPERS

In spite of the evidence of Christiaan Huygens's failed experiments with a marine timekeeper, there were still a few brave attempts to succeed with a mechanical clock design, especially once the incentive of the British rewards was announced. Until his death in 1727, Isaac Newton was requested to advise on submissions to the Board of Longitude, sent to him via the Admiralty. There is evidence that a number of applications were made and small sums of money were provided for experiments in the early years, long before the Board itself first formally met.

In his correspondence, Newton refers once to a clock design by 'The Quaker'. It is not known for certain who this is, but it is possibly a reference to the well-known Quaker clockmaker Daniel Quare (*c.*1649–1724), who is believed to have experimented with marine clock designs.

Among many published proposals for solving the longitude problem was one in the name of Jeremy Thacker, from Beverley in

An early attempt at a marine timekeeper by Henry Sully, 1724. This is one of three known clocks of this type. All failed because they relied on gravity to control the timekeeping.

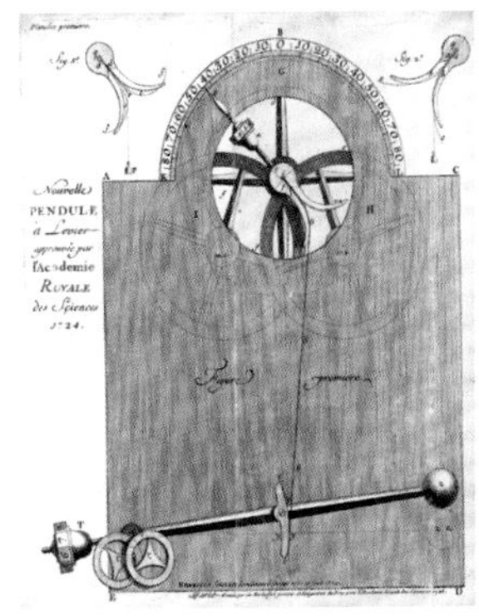

Engravings from Sully's published description of his marine timekeeper design, 1724.

East Yorkshire. The pamphlet, printed in London in 1714, contains many extremely interesting ideas for a marine clock. However, if such a clock was ever made it does not appear to have succeeded, as nothing more is heard of it, and the pamphlet is now thought to have been a spoof.

Determined efforts were also made by Henry Sully (1680–1728), an English clockmaker who worked for most of his life in France. An admirer and follower of Huygens's work, Sully made several attempts during the early 1720s to develop a workable mechanical design, producing and testing three prototype marine clocks. These designs, however, all sought to further develop timekeepers with gravity-controlled oscillators and, consequently, were destined to fail when employed at sea. Predictably, none of Sullly's experiments were successful and only served to confirm the widely held view, both in the scientific establishment and among the clockmaking fraternity, that a practical marine timekeeper was simply impossible.

It is remarkable then that not only was a solution to the longitude problem found, but also that it should prove to be a timekeeper after all. And what makes the story even more extraordinary is the nature of the man who made this technological breakthrough.

JOHN HARRISON, CLOCKMAKER AND SCIENTIST

John Harrison, a joiner from Lincolnshire, of a relatively humble background and with little formal education, took on the scientific and academic establishment of the day and by sheer determination, coupled with extraordinary innate technical insight, finally succeeded, qualifying for the full longitude reward and winning worldwide acclaim.

John Harrison, by Thomas King, c.1767. Harrison is shown holding a watch made to his designs by watchmaker John Jefferys and with H3 behind his right shoulder.

CHILDHOOD AND EDUCATION

Born on 24 March 1693 at Foulby, near Wakefield in Yorkshire, John Harrison was the eldest son of Henry Harrison, an estate carpenter, believed to have been employed by Sir Rowland Winn at Nostell Priory, next to Foulby. When John was about four years old, the family moved to the small village of Barrow, on the south bank of the Humber and in neighbouring Lincolnshire, just south of the large and rapidly expanding port of Hull. Here the Harrisons became established as notable members of the local community, with John's father soon taking up the role of Parish Clerk. They had four further children: two girls, Mary and Elizabeth, and two boys, Henry and James.

Little is known of John's early education, but this is likely to have been acquired at home, along with the woodworking skills taught to him by his father. Harrison was, nevertheless, a precocious youth, sufficiently bright to catch the attention of a visiting clergyman who, when John was a young man, lent him a copy of the celebrated lectures on mechanics by Nicolas Saunderson (1682–1739), Lucasian Professor of Mathematics at Cambridge. Harrison copied out the whole book. It remained a treasured possession and over the years he annotated it with copious notes. John also learned the art of land surveying, making his own plane table and compass, the portable instrument used in the field by surveyors of the period for making maps.

Harrison also displayed a great interest in music during his childhood, an interest both theoretical and practical, which was to stay with him throughout his life. He became an accomplished bass-viol player and was the choirmaster at the Barrow parish church. He also retuned the bells at Holy Trinity and St Mary Lowgate churches in Hull. In later life, Harrison published a radically different method for tuning the musical scale. He even understood that the high-frequency oscillations of musical instruments have implications for timekeeping design.

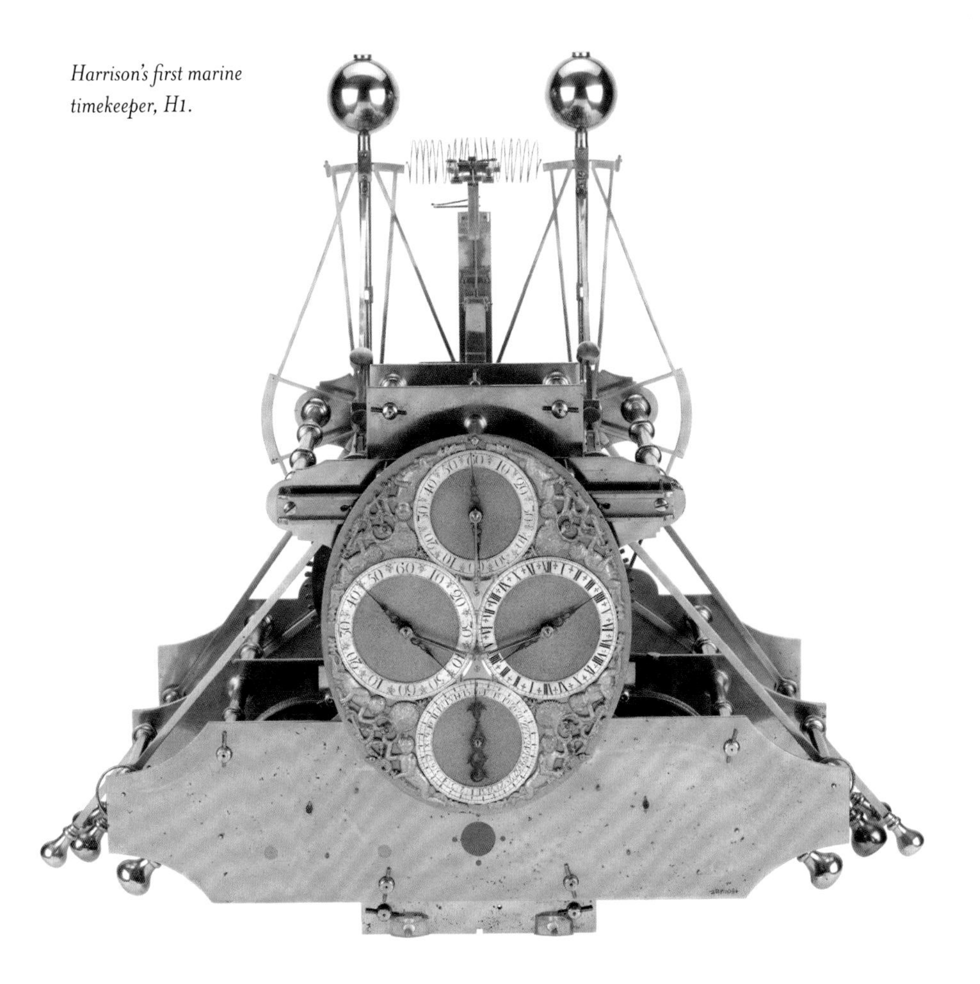

Harrison's first marine timekeeper, H1.

THE EARLY CLOCKS

Where Harrison's interest in clockmaking came from, we simply do not know. We do know that in 1713, at the age of 20, he had completed a longcase clock, which, while relatively ordinary in external appearance, has a mechanism made almost entirely of wood. Although this choice of material may seem logical enough for one trained as a joiner, making small clock mechanisms out of wood was unheard of in England. There were a few continental precedents emanating from southern Germany and one or two characteristics of Harrison's early work suggest the influence of Continental clockwork, so it is possible that he came across examples of these in the port of Hull.

Three of Harrison's early wooden clocks have survived. It is unlikely many more were made at this time and Harrison expert Andrew King has suggested these three clocks could have been made for members of Harrison's family. The first, with the movement signed and dated 1713, is now preserved in the Worshipful Company of Clockmakers' Museum, within the Science Museum, London. The second, similarly signed, and dated 1715, is now in the Science Museum's own collection. The third, dated 1717, is at Nostell Priory, though acquired in the early twentieth century and not directly as a result of Harrison's origins there.

The movement of Harrison's earliest surviving longcase clock, dating from 1713. Although made of oak and boxwood, the design is otherwise similar in principle to ordinary clocks of the day.

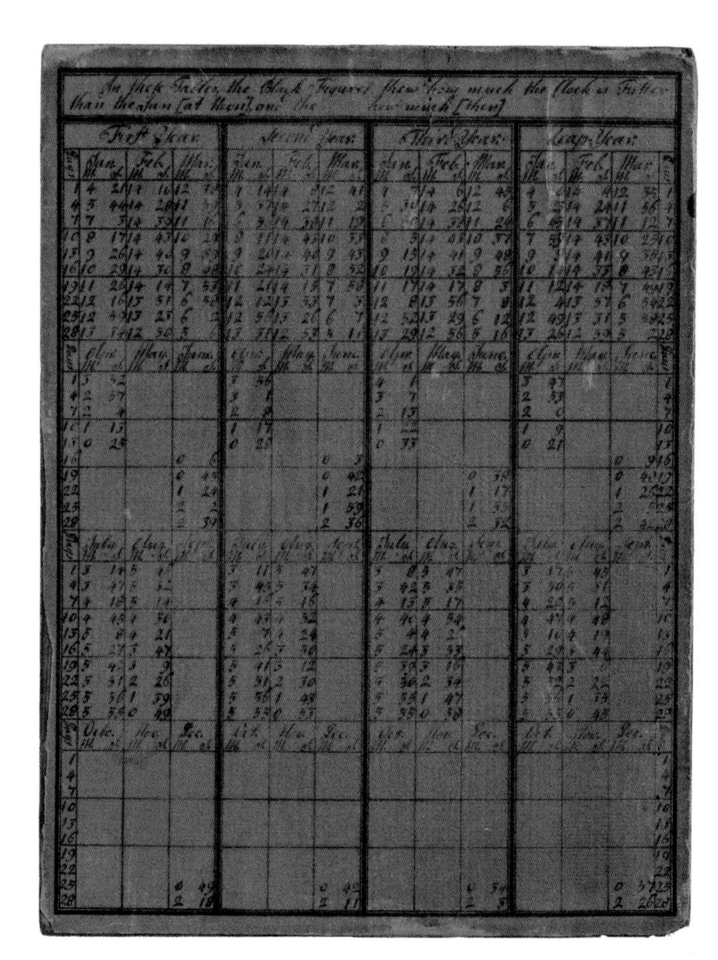

The 'Equation of Time' table, pasted to the inside door of Harrison's pendulum clock of 1717. This enabled the owner to set the clock using a sundial.

Simply constructed in oak, these clocks all have the conventional anchor escapement and the *countwheel* system of hour-striking, standard for longcase clocks at the time. The wheels and pinions are also made of wood, the wheels of oak and the pinions of boxwood (*Buxus sempervirens*). Even in these early clocks, Harrison's ingenuity is evident. In constructing his wheels, he used the mill-wright's technique of morticing segments of three or four teeth into the rim. This enabled him to cut the segments along the grain of the wood which, when inserted radially into the wheel, made the teeth considerably stronger. Having said this, in other respects the clocks are essentially wooden versions of contemporary metal clock designs and, with steel pivots and brass bushes in the wooden frame, these movements still required oiling in the usual way.

Today, none of these clocks has its original case, but a section of the trunk door of the 1717 clock does exist in the Clockmakers' Company collection. It provides an interesting insight into Harrison's education. Pasted to the piece of the door is a manuscript giving times of sunrise and sunset throughout the year, and an equation table. This table provides data for each day of the year, showing the difference between 'mean time' (uniform, clock time), and 'solar time' (time by a sundial, which varies throughout the year). The information can be used to set clocks correctly using a sundial. The table is in Harrison's own hand and suggests he may have had sufficient astronomical knowledge and skill to determine the data himself.

John had married in 1718, but sadly his wife Elizabeth died just eight years later. Within six months, Harrison got married again, to another Elizabeth. During the latter part of his early career as a clockmaker, John had his younger brother, James Harrison (1704–66), to help him and it seems that the two developed an interest in the technical aspects of clocks and watches. For example, the brothers began to time pocket watches in several different positions to study the effects of gravity on their timekeeping, a very early example of what became standard practice in precision watchmaking in the nineteenth and twentith centuries. Perhaps these experiments were associated with Harrison's early interest in winning the longitude reward, though Harrison states that he did not hear of it until 1726.

THE BROCKLESBY PARK CLOCK

The Harrison brothers' first major clockmaking project – and an important early commission for them – was a revolutionary turret clock made in the early 1720s for the stables at Brocklesby Park, the seat of the Pelham family (later the Earls Yarborough). The clock, which was also constructed almost entirely of oak, was groundbreaking because it needed no lubrication. Even modern clock oil tends to thicken with age, to creep away from where it is needed and to turn acidic or evaporate. Mostly deriving from animal or vegetable oils, eighteenth-century lubricants were particularly poor and tended to be one of the major causes of clocks failing to work.

With no formal education, John Harrison was always a radical thinker. Instead of worrying about ways to improve the oil, he designed a clock that didn't need it, the first of its kind and one of only a few ever to have been made. Harrison was particularly adept at selecting the best materials for an application and again employed boxwood, this time for the bearings in the clock, which, in combination with brass pivots, provided a very effective oil-free bearing surface. Later, he discovered that the dense, greasy, tropical hardwood lignum vitae (*Guaiacum officinale*), again in combination with brass, would perform even better.

The clock originally had an anchor escapement, but Harrison soon improved this with a new invention, the *grasshopper escapement*, which employed no sliding actions and therefore did not require oiling. The escapement was given this peculiar name because of the

The movement of the clock in the stables at Brocklesby Park, Lincolnshire, made by the Harrison brothers, c.1722. This revolutionary clock proved to be the foundation stone for John Harrison's later precision clock designs.

movement of its pallets (the two little wooden pieces that receive the pushes or impulses from the clock mechanism). To avoid sliding actions, which would need lubricating, these pallets are designed to 'jump' out of engagement after each impulse, their action being reminiscent of the hind legs of a grasshopper.

As well as being incredibly well designed, the Brocklesby Park clock is exceptionally and beautifully made, revealing Harrison's first-rate skills as a joiner and cabinet maker. More than 300 years after its construction, the clock is still at Brocklesby Park, continuing to run reliably and keep excellent time – and still without the need for lubrication.

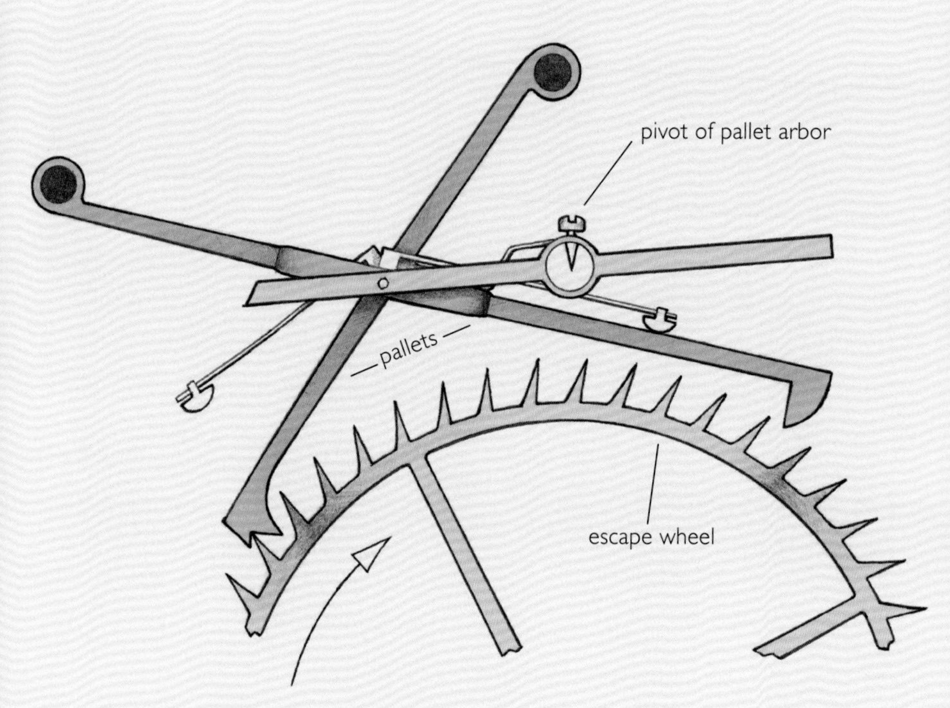

Harrison's grasshopper escapement, c.1725.

THE PRECISION PENDULUM CLOCKS

Flushed with this success, Harrison continued to develop more accurate and reliable clock designs. In the mid-1720s, under his brother's direction, James Harrison set to work on a series of remarkable precision pendulum clocks – smaller, longcase versions of the turret clock – to see just how far they could push the capabilities of this design. Superficially these clocks look like

The cottage on Barton Road, Barrow, to which Harrison moved in 1726. Sadly, it was demolished in 1968 while being considered for listed building status.

Harrison's early longcase clocks, but in detail they are very different indeed. Like the turret clock, they run without lubrication, having smaller versions of his grasshopper escapement and using lignum vitae bearings. The wheelwork is still of oak, but the pinions are made of little lignum vitae rollers, mounted on brass pins, so the wheel teeth are actually in rolling contact during meshing. Three of these precision longcase clocks have survived, being dated 1726, 1727 and 1728, the latter two on display in Leeds City Museum and the Clockmakers' Company Museum in London, respectively.

John Harrison's victory over the problem of lubrication by eliminating the problem itself was ingenious, but not in keeping with his usual scientific method. His usual approach was to accept the presence of an 'enemy' and negate the effect by compensating for it. Using this more typical method, he eliminated another significant error in these precision pendulum clocks: that caused by the effects of temperature change.

Clocks go slower when they get warmer because the pendulum rod expands and lengthens, and longer pendulums beat more slowly than shorter ones.

Harrison's first precision pendulum clock, made in 1726. This clock was one of a pair Harrison used as his time standard and also as a test bed for his precision clock development.

For it to keep time constantly, the pendulum's *effective length* must not change. The effective length is the distance between the point of suspension and the centre of gravity of the pendulum. Harrison solved the problem of temperature change by inventing a pendulum that, instead of having a simple rod, has a 'gridiron' made up of an alternating series of brass and steel rods, in which the downward expansion of the steel rods is counteracted by the upward expansion of the brass rods. In this brilliantly clever design, although all the rods are expanding, the effective length of the pendulum remains the same and it continues to keep time.

As a result, Harrison tells us that these early precision pendulum clocks achieved the astonishing accuracy of a variation of no greater than one second in a month, a performance far exceeding the best London clocks of the day. And because the clocks had no oil, they maintained their performance for much longer than conventional clocks. Harrison developed two of these clocks in his workshop in tandem, his ingenious scientific method being to use one of the clocks as a control while he made adjustments and improvements to the other, then switching the clocks and using the improved clock as a control while the other was adjusted. Both these clocks were then employed as regulators, to test his other clocks. He notes that he used the passage of stars across the sky at night in order to time them precisely. Accurate time was derived by observing the exact moment particular stars disappeared when aligned with the glazing bars of his window and a neighbour's chimney. This highly resourceful technique had been suggested by various authors, including Christiaan Huygens in the previous century, and was included in Saunderson's lectures, where Harrison may have learned of it. Using this system, Harrison was able to gauge time to the remarkable precision of a twentieth of a second.

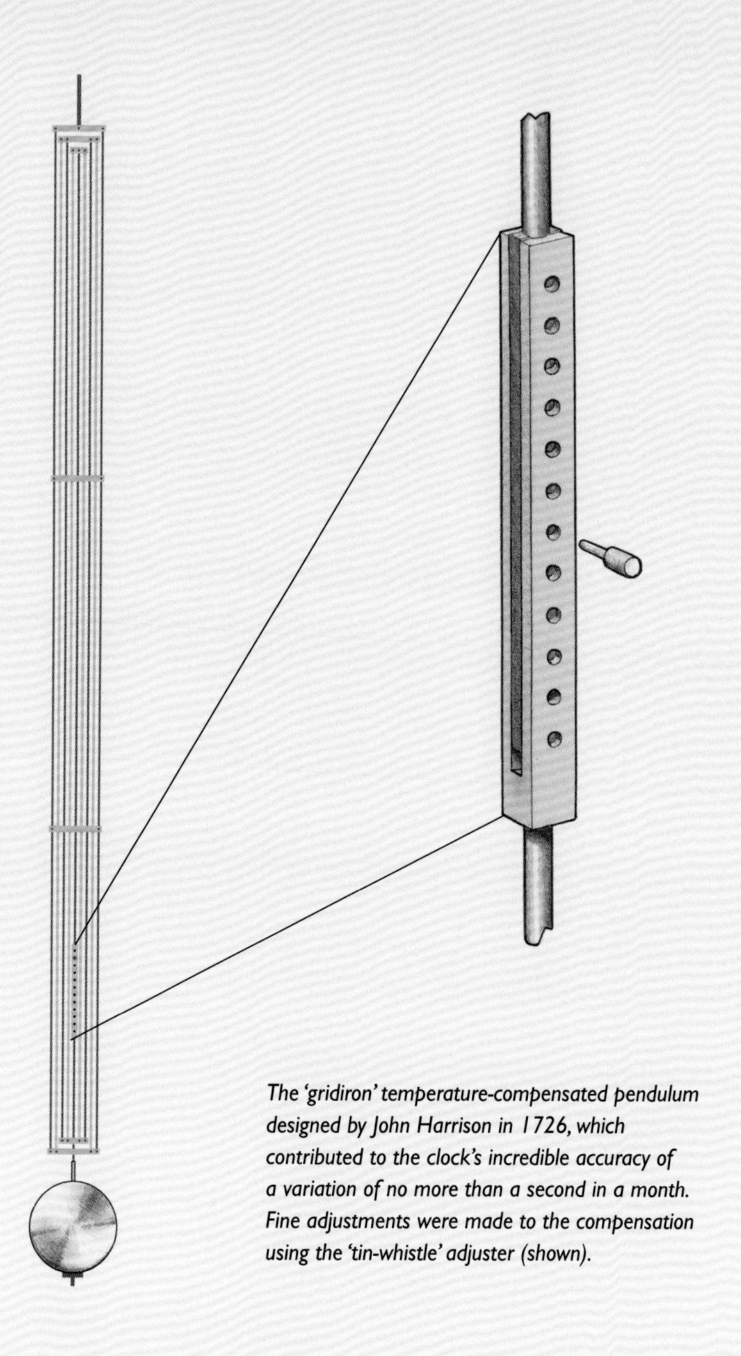

The 'gridiron' temperature-compensated pendulum designed by John Harrison in 1726, which contributed to the clock's incredible accuracy of a variation of no more than a second in a month. Fine adjustments were made to the compensation using the 'tin-whistle' adjuster (shown).

THE MARINE TIMEKEEPERS

A simple calculation based on the terms of the Longitude Act shows that, in order to qualify for the main longitude reward of £20,000, a timekeeper would have to keep time with a variation no greater than 2.8 seconds a day during the six-week trial voyage. Before 1750, the only commonly available portable timepieces, watches, were insufficiently accurate; even the very highest-quality watches of that period lost or gained at least a minute a day. The only timepieces capable of the required accuracy were large pendulum clocks, fixed rigidly to the wall like Harrison's regulators. So, given the options available to them, potential designers of marine timekeepers, such as Harrison, saw only one logical course of action: to win one of the longitude rewards, they would have to make a portable version of a precision pendulum clock. As we shall see, however, this progressive and apparently logical approach was not the correct one, but no one realised this at the time.

Over the following few years, Harrison therefore formulated a plan for a large marine timekeeper and it is recorded by Harrison's grandson that he visited London in 1727 to seek support to make it. Harrison himself tells us that he came south at about this time to seek funding and moral support, taking with him drawings and a written description of his proposal for the timekeeper.

It has been suggested by Harrison's biographer, Humphrey Quill (1897–1987), that this visit was more likely to have been two or three years later, in 1730, as a manuscript description of his first marine timekeeper, dated that year, has survived. 'The 1730 manuscript', as this little book is known, is now preserved in the Clockmakers' Company collection. However, it is more likely that Harrison did come to London in 1727 with preliminary sketches and that the 1730 manuscript, which is very carefully worded and illustrated, is a refined version, the result of further development and advice, and possibly intended as a publication for use on a follow-up visit to London.

Harrison records that on his first visit he went initially to Greenwich to seek the advice of the Astronomer Royal. This was Edmond Halley (*c.*1656–1742), of comet fame, who received Harrison very kindly. At the time, Halley was despairing of ever completing the 'lunar tables' that were to contain the data required for the lunar-distance method.

Halley, however, not being in any way horologically qualified, felt unable to judge the soundness of the plans and suggested that Harrison go directly up to London to see George Graham (*c.*1674–1751). Graham (who told Harrison of Halley's troubles) had been partner to Thomas Tompion, and was then the greatest and most highly respected maker of watches, clocks and instruments. Halley therefore warned Harrison to be brief and to the point when meeting Graham. It is clear that Harrison was not the best at expressing himself succinctly, and, what with making his clocks from wood and being based in the wilds of Lincolnshire, one can understand that someone as busy and important as Graham may initially have been prejudiced in his assessment of the man with whom he was dealing. Harrison writes of Halley:

> *[He] caution'd me how to begin with Mr Graham […] in as few words as possible to let him to understand that I had indeed something worthy [of] notice to communicate to him; but as notwithstanding that piece of*

Left: *Astronomer Royal, Edmond Halley, by Sir Godfrey Kneller, c.1720. Halley encouraged Harrison and recommended he consult George Graham.*

Right: *The watchmaker George Graham, by Johan Faber the Younger after Thomas Hudson, c.1740. Graham gave Harrison advice and moral support, as well as some money to get him started in making H1.*

> *advice and my doing my best [...] Mr Graham began, as I thought it very roughly with me [...] which [...] occasion'd me to become rough too; but however, we got the ice broke [...] and indeed, he became as at last, vastly surpris'd at the Thoughts or Methods I had taken.*

Harrison, having arrived at 10 a.m., was still discussing timekeeper design with Graham at dinner in Graham's house late into the evening. Graham even extended the offer of a loan to support Harrison's work: a greater demonstration of confidence could not be imagined. It is interesting to speculate at what point the 'ice broke' in their discussion and which particular 'thoughts or methods' impressed the great watchmaker. It is known that, since at least 1715, Graham himself had been trying to design a pendulum with temperature compensation using brass and steel rods, but he had not been able to work out how to do it; in the end he came up with another system, using mercury in a glass jar as the bob of the pendulum. No doubt when Harrison showed him how he had designed his type of compensation pendulum, Graham realised this was no ordinary carpenter from the country.

On returning to Barrow, Harrison spent the next five years or so, along with his brother James, constructing the extraordinary timekeeper known today as 'H1'. After preliminary but rigorous tests on a barge on the River Humber, Harrison felt ready to have the machine tested more formally. H1 was brought to Graham in London and publicly displayed to the scientific community. It became quite a celebrity, hugely impressing all who inspected it, and was widely regarded as one of the wonders of the age. Harrison was besieged with requests to see the timekeeper by both scientists and socialites, and was very happy to show his creation to visitors, explaining the various parts and describing their function. It is also possible that some came to see Harrison the man: the curiosity from the country, the ingenious clockmaker.

In 1735, the Royal Society (whose members included Graham and Halley) issued a certificate of H1's great potential, which was then used to persuade the authorities to grant the new timekeeper some form of trial. This, until his death eight years before, would have fallen to Isaac Netwon to consider, but now it was up to the Admiralty's other advisers to recommend to the First Lord of the Admiralty that Harrison and his timekeeper should be given a semi-official trial. And so it was that H1 and its maker sailed, on board the warship *Centurion*, to Lisbon in May 1736.

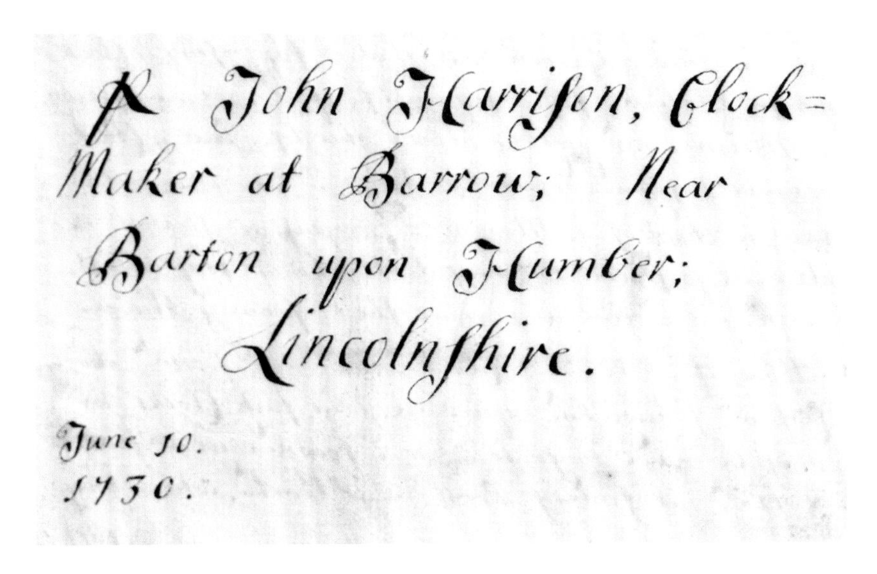
ꝑ John Harrison, Clock=
Maker at Barrow; Near
Barton upon Humber;
Lincolnshire.

June 10.
1730.

Harrison's signature at the end of the 1730 manuscript. This document summarised Harrison's proposals for his first marine timekeeper, H1.

Before departing, the ship's captain, George Proctor (d.1736), wrote that he found Harrison 'a very sober, very industrious and withal, a very modest man' but he feared that at sea 'the difficulty of measuring time truly [...] gives me concern for the honest man and makes me fear he has attempted impossibilities'. After a week-long voyage in tempestuous weather, Proctor wrote from Lisbon that Harrison '...was sea sick withal, but [...] seems satisfied that the motion of the ship was not in the least detrimental to its keeping true time'. The ship's log, however, reveals H1 not performing as well as it might on the outward voyage and some of the positions calculated from its readings appear to be considerably wide of the mark. In Lisbon, H1 was transferred to the ship *Orford* and does appear to have functioned much better on the long homeward voyage. Indeed, Harrison used H1 to correct a misreading of the ship's longitude and prevented what could have been a very serious incident. All the officers believed that the first land sighted by the ship was 'The Start' (Start Point near Dartmouth). Harrison and H1 proved that it was, in reality, 'The Lizard' (Lizard Point on the Penzance peninsula). This is located nearly 60 miles to the south-west – it is the southernmost point in England – and revealed to the crew that the ship was in fact in a much more perilous position, enabling evasive action to be taken in the nick of time. In due course, in recognition of this significant feat, the Master of the ship, Roger Wills, presented Harrison with a certificate outlining these important contributions.

On 30 June 1737, the Board of Longitude was convened officially, for the first time, to hear how H1's trial had gone and to inspect this model of 'high technology'. The news the Board received was evidence, after all, that a marine timekeeper might just prove to be a practical solution to the longitude problem.

'H1', THE FIRST WORKABLE MARINE TIMEKEEPER

Today, 'timekeepers' are referred to as *marine chronometers*, but the term chronometer was not generally used until after Harrison's death. The word 'timekeeper', however, had very special significance; it was only used to describe a portable machine capable of high accuracy. It should also be noted that the 'H' abbreviations, used today to refer to Harrison's timekeepers, are a relatively recent denomination, first applied by Lieutenant-Commander Rupert T. Gould in the 1940s during his restoration work on them.

As mentioned, Harrison designed his first marine timekeeper to be, conceptually, a 'portable' version of his precision pendulum clocks, and it is important to bear this in mind if one is to understand H1 properly. Unlike the pendulum clocks, which go for a week with one winding, all Harrison's timekeepers run for one day only, but H1 has wheelwork of oak with lignum vitae roller pinions. The main frame and ancillaries of the timekeeper are all made in brass or other alloys, where possible avoiding the use of steel, which would rust at sea and be affected by magnetism. Instead, Harrison used two types of bronze, a low-tin bronze where good tensile strength was required, and a high-tin bronze where high compression strength was needed. The major difference between H1 and the pendulum clocks is that H1 does not require gravity for any of its operations – an essential prerequisite for a marine timekeeper, but one which many earlier designers, including Huygens and Sully, had not fully understood. Thus, the timekeeper is spring-driven, with a fusee to ensure a uniform driving force. This fusee is, however, most unusual in that it employs two chains and two barrels, positioned at 180° to one another, hugely reducing the load on the pivots of the fusee.

In spite of its relatively good performance on the *Orford*, H1 did not perform nearly well enough to win even the smallest of the longitude rewards. Even before its trial Harrison probably knew the design could be improved; it was, after all, very much a prototype and one on which the Harrisons never even put their signatures. Wishing to move on, Harrison didn't ask for a second, official trial of H1, but requested financial assistance from the Board to make a new version of the timekeeper. Nothing like H1 had ever been seen before and the Commissioners were undoubtedly very impressed both by the machine and by its inventor. They allocated him £250 there and then, with the promise of another £250 on completion of an approved machine. In agreeing to this support, the Commissioners were, in effect, instigating the very first government-sponsored research and development project undertaken by a private contractor.

On his visits to London, and especially once he had been introduced to George Graham's circle, Harrison discovered how the city's unique horological facilities and connections made work easier. Almost anything he needed in the way of horological services or materials could be sourced in London. There is no doubt either that Graham and his circle would have been able to advise Harrison on written scientific and horological sources (though there were few of the latter in English at the time), and we know Graham himself discussed and advised Harrison on some details of his timekeepers (advice that was not always taken).

Harrison now decided that, if he were to succeed in creating a marine timekeeper, he would need the support of the London trade and he moved to the capital in 1736, soon after his return from

Lisbon. His first home was in Leather Lane, in Holborn, but in about 1739 he moved west to Red Lion Square, where he would remain for the rest of his life. It seems that James accompanied his brother for the first year or so, but there may have been a breakdown in the partnership as James did not stay long; by 1738 he was certainly living back in Lincolnshire.

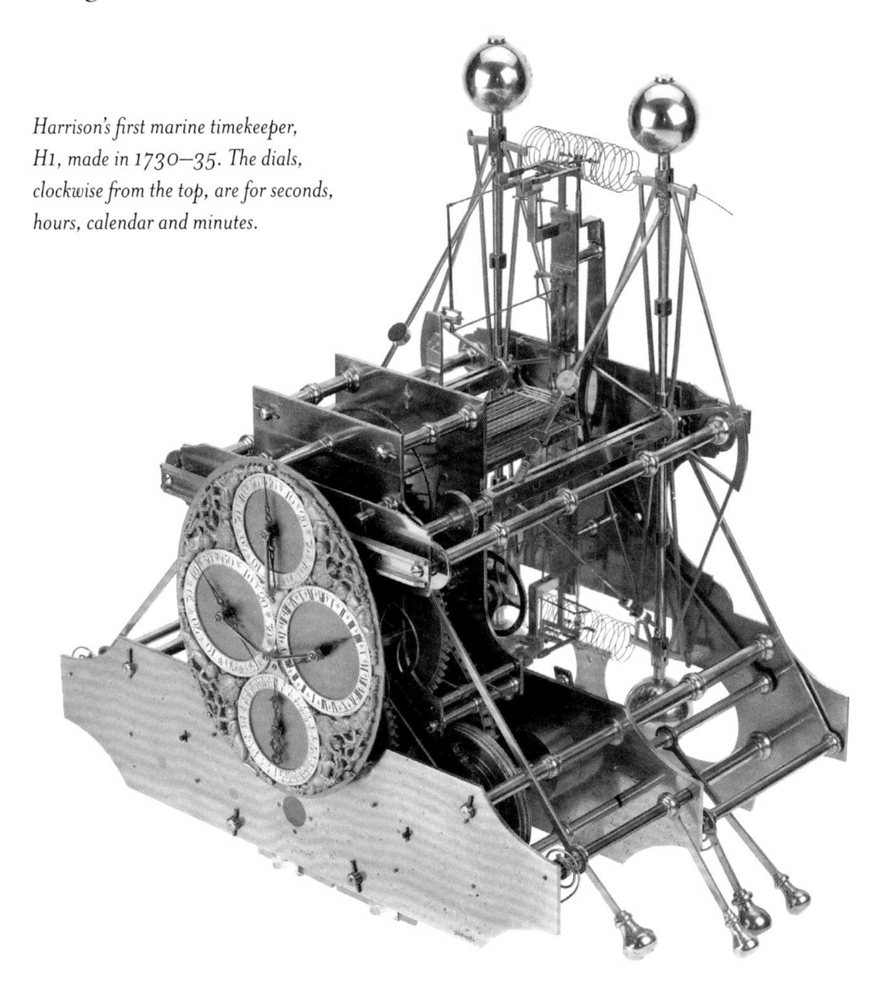

Harrison's first marine timekeeper, H1, made in 1730–35. The dials, clockwise from the top, are for seconds, hours, calendar and minutes.

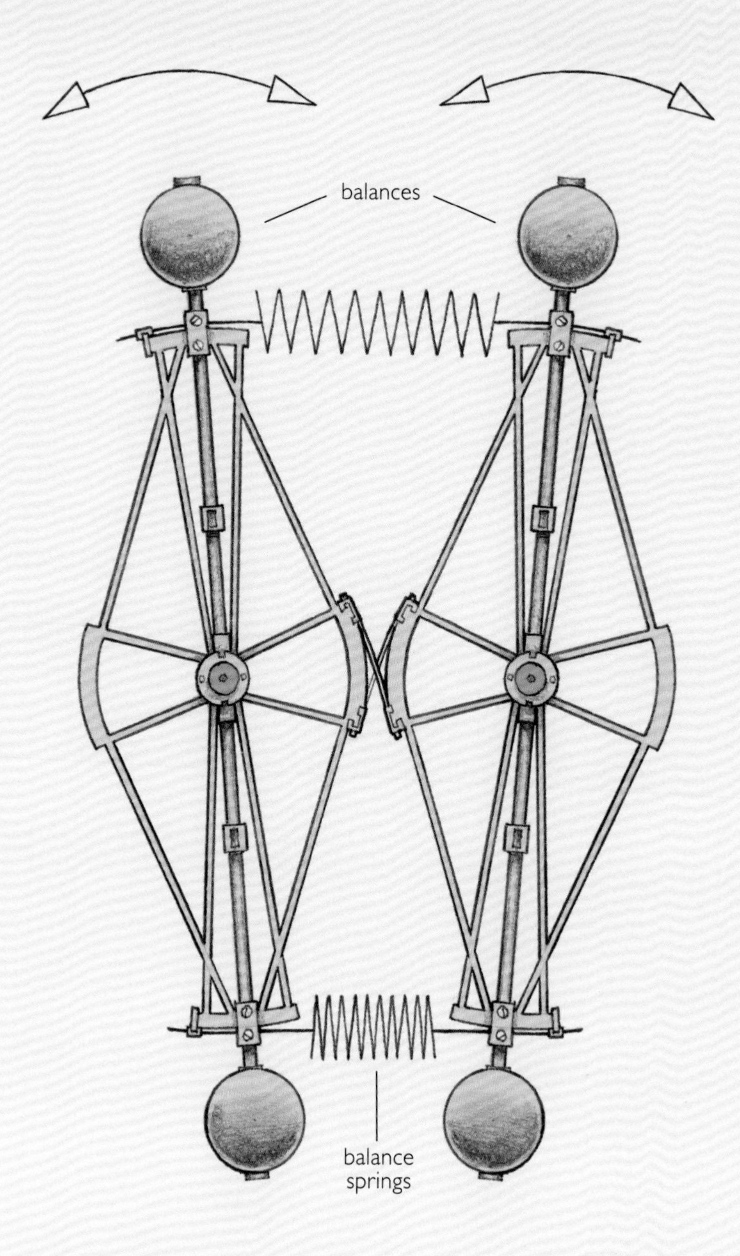

H1's balances are linked at the centre by cross-over ribbons. Harrison believed they would therefore be unaffected by external motion.

THE TECHNOLOGY OF H1

MAINTAINING POWER

Incorporated within the fusee of H1 is one of Harrison's inventions, which would prove of lasting benefit to chronometer- and watchmakers. This was his form of the mechanism known as ***maintaining power***. In weight-driven clocks, and in spring-driven clocks with a fusee, when one winds up the mechanism, the drive to the wheels is temporarily removed and the clock may lose time, or even stop working. To get round this, the mechanism is fitted with maintaining power, which ensures a continuing drive during winding. The pre-Harrison version of this mechanism had to be actuated manually, but Harrison's type is automatic and one is unaware of its operation during winding. By the late eighteenth century virtually all precision clocks, watches and chronometers employed this design.

TWIN BALANCES

The movement of H1 is considerably larger than the wooden longcase clock movements, weighing 34 kg and standing 63 cm high. For the timekeeping element, Harrison replaced the pendulum with two interlinked bar-balances (like a pair of dumbbells), connected across their centres by cross–over ribbons (a type of frictionless gearing). The balances swing in anti-phase (i.e. towards and away from each other) and the interconnection between them was supposed to ensure that any external motion affecting one balance was counteracted by that same effect on the other balance. It is true that if the machine is turned around while running it does appear 'not to notice' that it is being moved. The balances are also linked with helical steel springs at top and bottom, to provide a restoring force, like gravity acting on a pendulum (like Hooke before him, Harrison referred to these springs as providing his 'artificial gravity'). So, when swung apart, the balances oscillate together, each swing taking one second.

OIL-FREE

The balances themselves do not run directly in the frame of the clock, but sit and roll on special 'anti-friction' segments, to render their action almost frictionless. Thus, with an adapted version of the grasshopper escapement and the other elaborate anti-friction devices kept in the design, the clock

required no lubrication. Harrison also adapted the gridiron principle from his pendulum in order to compensate for changes in temperature. In balance-controlled timekeepers, a rise in temperature not only causes the balance to get larger, which will make it swing more slowly, but a higher temperature will also cause the balance springs to become weaker, reducing the 'artificial gravity' and making the slowing effect even greater.

FIRST COMPENSATION BALANCE

An important technical point about H1's temperature compensation, Harrison realised, was that to perform at its best, just as in his pendulum clocks, the compensating mechanism should be built into the balances themselves and they were originally constructed that way. Unfortunately, he found that this first pioneering design wasn't sufficiently reliable in operation and he was forced to rearrange the gridirons in a fixed position within the frame of the clock. Nevertheless, in later years he still maintained that the compensation would be best within the balance and after his death the principle of a compensation balance was accepted as the norm for all precision watches and chronometers.

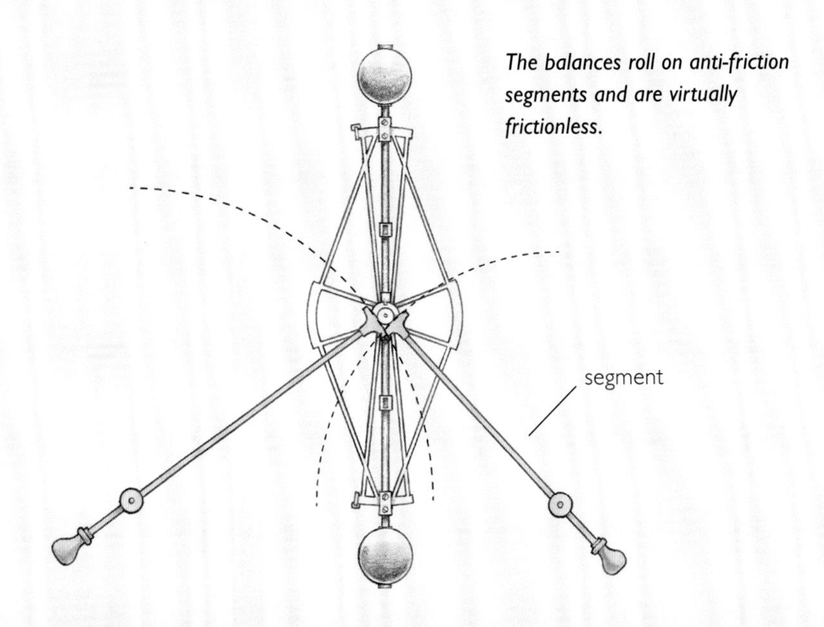

The balances roll on anti-friction segments and are virtually frictionless.

Opposite: *The mechanism of H1.*

THE SECOND TIMEKEEPER, 'H2'

Harrison began work on the second timekeeper immediately and, by January 1741, he was before the Board again. However, by this time he had already perceived a deficiency in the design of H2 and had begun work on a third timekeeper, H3. Notwithstanding this, H2 is as remarkable a timekeeper as H1 and has a more professional feel to its construction. Harrison told the Board in a later interview that he had employed the services of a number of London tradesmen; this would certainly have been for the supply of materials such as brass plate and steel springs, and services such as engraving. In all probability he also employed workmen for tasks such as basic finishing, although the main design and layout of the movement would have been all his own, the details of which would have been carefully concealed from prying eyes until presented to the Board.

Larger and heavier than H1, H2 stands 66 cm high, weighs over 39 kg and is made almost entirely of brass. The only wood in the timekeeper is in the lignum vitae parts and the pallets of the escapement. The concept is fundamentally the same as H1's, except that the temperature compensation is of a simplified design and Harrison fitted a *remontoire* to H2.

As with H1, H2 would originally have been cased and mounted in large gimbals to ensure it remained horizontal at all times. The mounting itself does not survive but its form has been recorded in a small pen-and-ink colour-wash sketch by artist John Charnock (1756–1807), drawn at Greenwich about 1780 and showing H2 in its complete state.

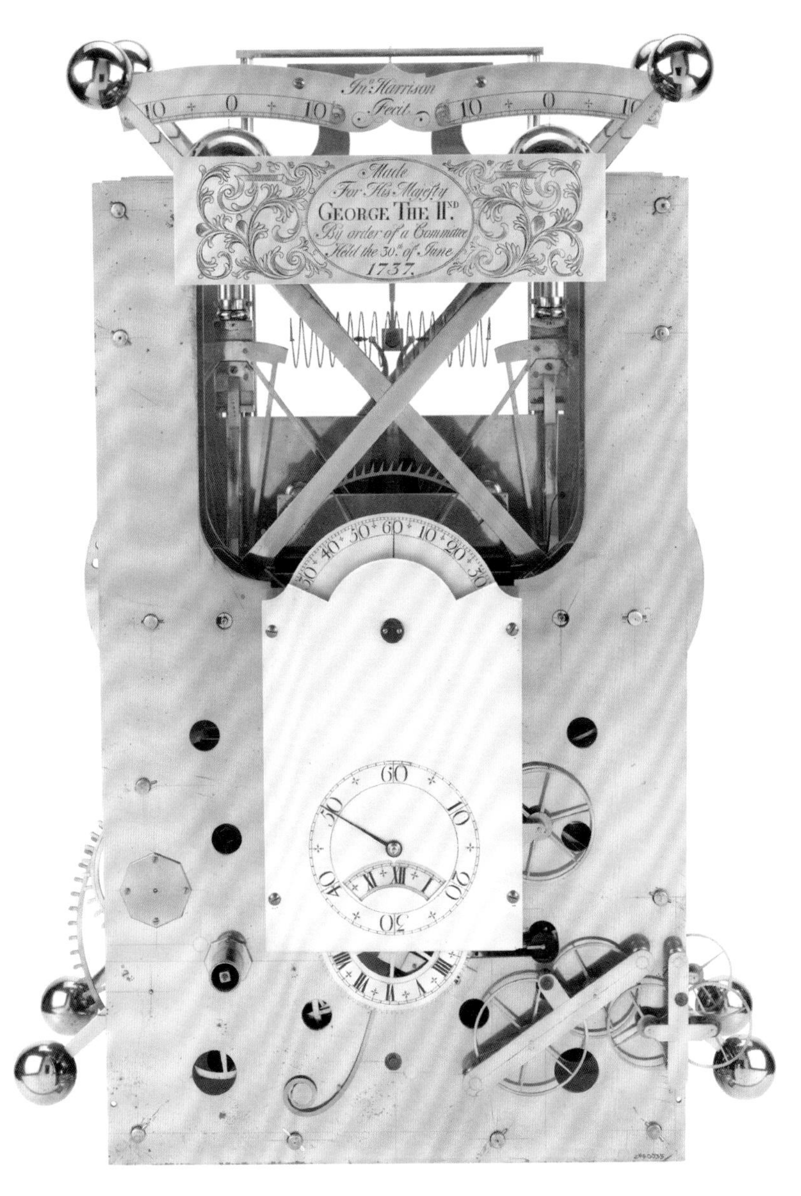

H2 took Harrison, probably with the help of London craftsmen, just two years to make – from 1737 to 1739. It is the heaviest of the timekeepers.

Left: *This tiny colour-wash drawing of H2 was done by the artist John Charnock, c.1780. It shows the original gimballed mount, now missing.*

Opposite: *The mechanism of H2. Part of the remontoire is visible in the centre of the image.*

THE REMONTOIRE

In an ordinary clock, which has a long series of wheels to supply energy to the escapement, small errors in manufacture of the wheels cause variations in the driving force. This, in turn, causes variation in timekeeping. The remontoire is intended to remove these variations.

A small spring directly drives the escapement, independently of the main wheel train. The function of the main train was now simply to wind up this small spring at regular intervals, as and when it needed it. In H2, this rewinding occurs every 3 minutes 45 seconds.

Harrison noted a deficiency in the linked bar balances when H2 was moved. As in H1, the cross-linking of H2's balances was supposed to render them insensible to external motions while running. But he discovered, very much to his consternation, that the timekeeper was somewhat affected by movement, owing to centrifugal force acting on the balances. If subjected to circular motion in a horizontal plane, the timekeeper would tend to go slow; if subjected to circular motion in the plane of the side elevation, the timekeeper would tend to gain.

He realised that the reason for this, in the first two types of motion, was that the balances were in the form of dumbbells; had he made them in the form of wheels, the problem would be removed. But there was no room in H2's design to convert to wheel balances, so, after more than two years of hard work and considerable expense, he was obliged to set H2 aside and start again. Any official trial must give him the best possible results: he may not be given another chance and he could not contemplate having H2 tested once he discovered this fault. Harrison's backers stayed with him and petitioned the Board on his behalf for more money to continue with H3. Still highly impressed with his ingenuity, the Commissioners duly awarded Harrison another £500 and work continued on the nascent third timekeeper.

THE MEANING OF THE ACT OF PARLIAMENT CONFIRMED

The record of this decision in the minutes of the Board of Longitude is especially interesting as it confirms the Board's view that just one successful West-Indies trial of a single timekeeper, as specified in the 1714 Act of Parliament, would qualify the maker for the full Longitude reward. This was a subject that in years to come would be denied by later members of the Board and would lead to conflict with Harrison and his supporters.

In wishing to make it clear to Harrison that the award of £500 must be considered part of the full reward if he later qualified for it, the Board minutes state (author's emphasis in bold):

> *...if upon Experiment* **[the West Indies trial]** *it shall be found that the said Machine will contribute to the finding out of the Longitude* ***& that thereby*** *the said Mr Harrison* ***shall be entitled*** *to receive any of the Rewards mentioned in the Act of the 12th of Queen Anne, that then, & in such case, the sum of £500 hereby granted to him & such other sum or sums of money as he hath heretofore received on account of other Engines invented by him for finding out the Longitude shall be deducted out of such Reward...*

No doubt then that in 1741 the Board of Longitude was of the opinion that one successful trial with one timekeeper was all that was required for the full reward to be granted and, naturally enough, that is what Harrison understood too.

Anno Regni

A N N Æ

R E G I N Æ

Magnæ Britanniæ, Franciæ, & Hiberniæ,

DUODECIMO.

At the Parliament Summoned to be Held at *Weſtmin-ſter*, the Twelfth Day of *November*, *Anno Dom.* 1713. In the Twelfth Year of the Reign of our Sovereign Lady *ANNE*, by the Grace of God, of *Great Britain*, *France*, and *Ireland*, Queen, Defender of the Faith, &c.

And by ſeveral Writs of Prorogation Begun and Holden on the Sixteenth Day of *February*, 1713. Being the Firſt Seſſion of this preſent Parliament.

LONDON,

Printed by *John Baskett*, Printer to the Queens moſt Excellent Majeſty, And by the Aſſigns of *Thomas Newcomb*, and *Henry Hills*, deceas'd. 1714.

The 1714 Longitude Act offered £20,000 for a method of finding longitude to within half a degree. Smaller sums were offered for less accurate methods.

HARRISON'S THIRD TIMEKEEPER, 'H3'

Unfortunately for Harrison, H3 was to be even more problematic than H2. Within five years it was running and under test, but from the outset it was clear that getting this design to keep close time would be difficult and Harrison was obliged to make constant changes to it. Years went by and although many improvements were made, he just could not get this timekeeper to perform to his expectations. Even after an astonishing 19 years of painstaking labour, H3 was stubbornly refusing to keep time well enough and, while Harrison learned a great deal from this herculean endeavour, its ultimate role was solely to convince him that the solution lay in another design altogether. So, after his initial success with H1, the 1740s and early 1750s must have proved something of a mid-life crisis.

However, none of this was known to the Board of Longitude, which continued to support him with grants until 1760, awarding him over £3,000 during the period. His supporters and other members of the scientific community were also unaware of the great difficulties Harrison faced and were still pinning their hopes on H3's success. Indeed, aware of the much wider implications for science of what Harrison was attempting, and even though H3 had still not provided the breakthrough, in 1749 the Royal Society awarded Harrison its highest honour, the Copley Medal, for his research work.

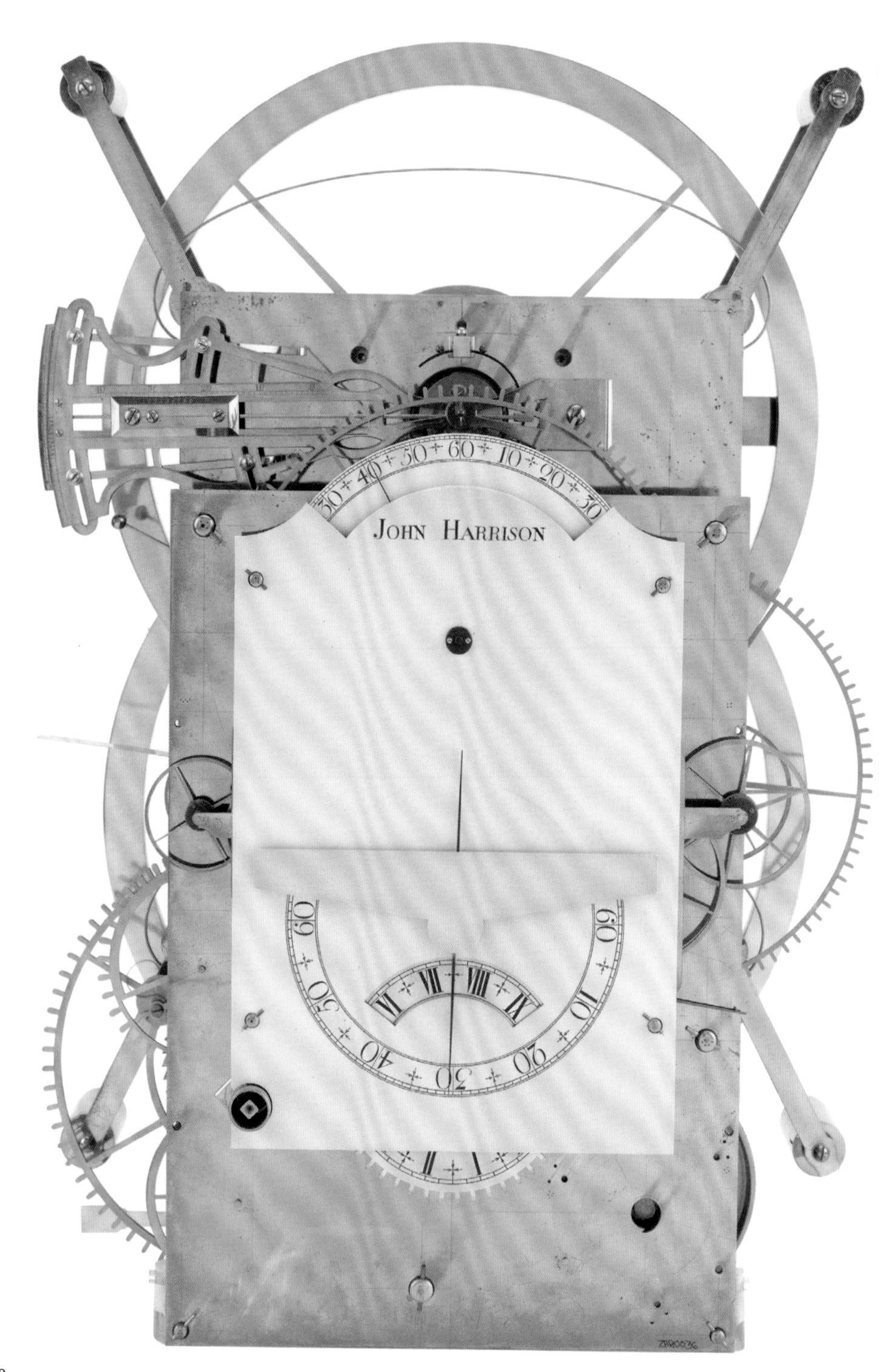
JOHN HARRISON

posite: *H3, rrison's third rine timekeeper, k 19 years to struct and adjust 40–59), though imekeeping never te came up to ectations.*

ght: *The chanism of H3.*

H3 itself, which was supposed to be Harrison's magnum opus, must have been the greatest technical disappointment of his life. In later years he could only refer to it wryly as 'my curious third machine', and his perplexed notes on this device represent the only example on record of the great man admitting to failure. Nevertheless, H3 is an extraordinary mechanism and contains inventions that would prove exceedingly important in the history of technology. The timekeeper stands 59 cm high and weighs 27 kg (43 kg in its case). The balances are wheels instead of dumbbells and are arranged one above the other. They are still linked together with cross-wires, beat seconds and are driven by the grasshopper escapement as with earlier machines. For his 'artificial gravity', instead of helical springs Harrison fitted one short spiral spring, which controls the upper balance only; a 30-second remontoire was fitted to provide uniform power supply; and the mechanism runs without lubrication as with the previous timekeepers. H3 was mounted like H2, in a glazed brass case, and is depicted behind Harrison in the portrait by Thomas King (d. *c.* 1769) (see pages 32–3).

The glazed case for Harrison's third timekeeper, H3.

THE BIMETALLIC STRIP

In order to compensate for temperature changes in H3, Harrison created a new and revolutionary device, the ***bimetallic strip***. This consisted of two flat strips, one of brass, one of steel, riveted together. Because brass expands more than steel with a rise in temperature, this bimetal bends into a curved shape with the brass on the convex side, and will bend back the other way if the temperature falls. With one end fixed, the movement of the other end of this device can then be used to adjust the balance spring automatically with changes in temperature. A shorter spring is normally stiffer, so, with a rise in temperature, the bimetal shortens the balance spring slightly, compensating exactly for the effects of temperature. The bimetallic strip was a tremendously versatile invention and would come into its own in the twentieth century. Most households today use the bimetallic strip as a thermostatic control in central heating, electric kettles, irons, toasters and so on.

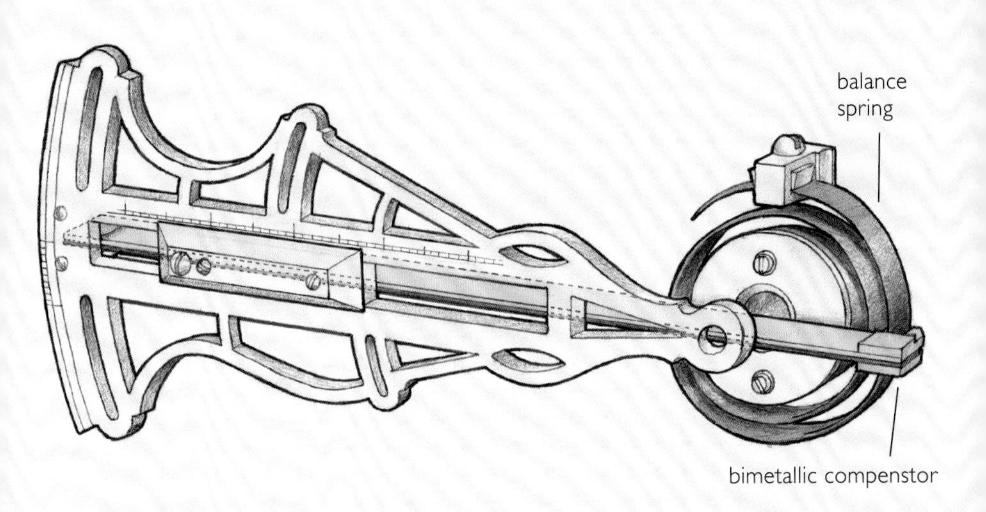

H3's improved temperature compensation was achieved with the bimetallic strip, another of Harrison's brilliant inventions that is still very much in use today.

THE CAGED ROLLER BEARING

One of the caged roller bearings from H3. In the nineteenth century, this seminal design of Harrison's would develop into the caged ball bearing and was of immense technological importance.

Another important invention that Harrison created for H3 was the ***caged roller bearing***, the ultimate evolution of his anti-friction devices. The central pivot of the bearing is surrounded by a series of four bronze rollers, mounted in a circle, in a little brass cage. These rollers run round inside a circular bronze track. When it is revolving, the pivot can take loading in any direction and the rollers will support it, in rolling contact. The whole assembly is virtually frictionless and requires no lubrication. In effect, it is the predecessor of the caged ball bearing, a device used in virtually every complex machine made today.

ISOCHRONISM

One of the biggest problems Harrison had with H3 was ensuring that the swings of the balances were ***isochronous***. This complex-sounding word (pronounced 'I sock ronnus') in fact describes a very simple concept. When a balance or pendulum is isochronous it simply means that all the swings, whether large or small, take the same time. Therefore, if the drive to the balances should occasionally be a little reduced, which would make the swings of the balances a little smaller, the timekeeping would still be correct. The balances of H3, however, were not isochronous, and Harrison spent many years trying to understand why and finding ways to correct the problem. In fact, the main problem lay in his use of the single, short spiral balance spring. But at this period, the complex science of spiral balance springs, which would be central to the later development of precision watches in the nineteenth century, was still largely a mystery. It is not surprising, then, that Harrison had such difficulties, but there is no doubt that he learned a great deal about balances and springs from his struggles with H3, so his 19 years were not entirely wasted.

OTHER WORK BY HARRISON

We know very little about Harrison's other activities during his life in London. Everything seems to have revolved around his work on the timekeepers and it seems highly unlikely that he, or his wife Elizabeth, had much time for socialising at any stage of their lives. We do know that Harrison's interest in music remained as strong as ever and that he continued to advise on the tuning of musical instruments and bells. He also occasionally took on private horological commissions. For example, in 1755 he designed a new form of anchor escapement for the turret clock at Trinity College, Cambridge, for which he was paid 3 guineas. The escapement was actually made for Trinity by another clockmaker, William Smith of Upper Moorfields (now Finsbury Square) in London. Unfortunately, the clock no longer exists (except for its setting dial), but there are at least four surviving examples of Harrison's escapement in other clocks made by Smith, and all still working, in England.

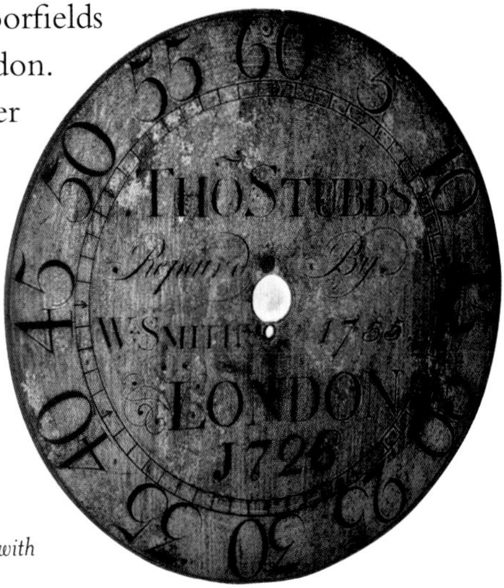

The setting dial from the clock at Trinity College, Cambridge. This is all that survives of the first clock with Harrison's anchor escapement of 1755.

8 6 4 0 4 6 8

Above: *The Harrison-type anchor escapement from one of four surviving turret clocks made by William Smith* c.1760.

Opposite: *Drawing for a new form of anchor escapement, designed by Harrison for the turret clock at Trinity College, Cambridge, in 1755. He was paid 3 guineas for it.*

THE R.A.S. REGULATOR AND A MODERN TRIAL

In parallel with his work on H3, from 1740 Harrison was also busy working on an equivalent regulator, to supersede his wooden-movement precision pendulum clocks, by then nearly 15 years old. This clock, now known as the R.A.S. regulator (the clock belongs to the Royal Astronomical Society), was, as far as we know, never finally adjusted by Harrison. Creating it was, in one sense, like the work on H3 in that it involved a long process of design and adjustment. In every other sense it was totally unlike H3 because this design was highly successful. So good was the performance Harrison was getting from it that he predicted that one day, once finally adjusted, this regulator would be able to keep time with variations no greater than 1 second in 100 days, a feat not achieved until William H. Shortt (1881–1971) introduced his 'free pendulum' clock, with a pendulum swinging in a partial vacuum, in the early 1920s. Since Harrison's day, his assertion has generally been dismissed as wishful thinking by commentators, as it was considered almost impossible to achieve such timekeeping stability with a pendulum swinging in air.

However, modern experiments with clocks made by the English horologist Martin Burgess of a design based on Harrison's original technology, have shown that, given time, Harrison would surely have proved his claim. The Burgess clock known today as 'Clock B', completed by Charles Frodsham & Co and adjusted and tested at the Royal Observatory in Greenwich between 2012 and 2016, performed with astonishing accuracy. Between April 2014 and April

The R.A.S. (Royal Astronomical Society) regulator. Made by Harrison in parallel with H3, this fixed, pendulum-controlled clock was Harrison's last word in regulator design. He predicted a variation of no more than 1 second in 100 days from it.

2016 (over 700 days), the maximum deviation from correct time was just over 2 seconds, at the end of 2015. In the following months this error then reduced again and by the end of two years running the clock was precisely on time, with no error showing at all. Had Harrison completed his adjustments on the R.A.S. clock and made the design generally available, eighteenth-century scientists could have had a time standard far beyond anything available to them at the time.

It was developing a successful marine timekeeper that was Harrison's principal concern, however, and by the early 1750s, while continuing his struggles and failures with H3, he was doing what all good inventors do and looking elsewhere to see if there wasn't another way to crack the problem.

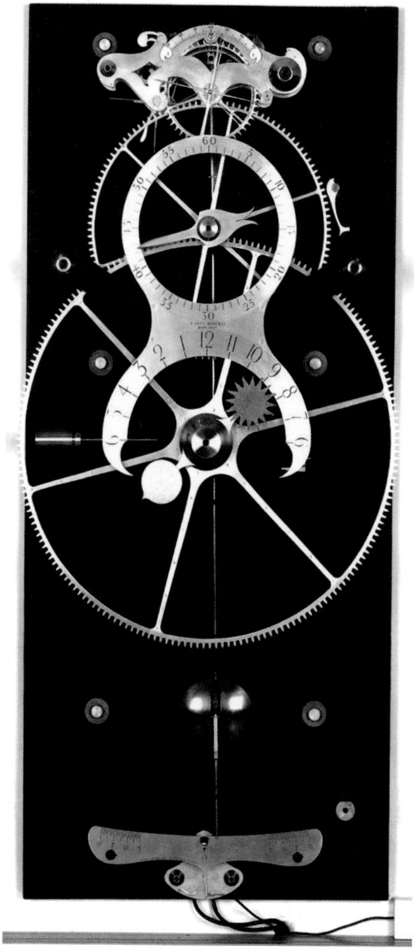

'Clock B', the clock that has proved Harrison's pendulum technology is capable of keeping time to within 1 second in 100 days.

THE BREAKTHROUGH

Among the small number of horologists and scientists who still believed a marine timekeeper was a possibility, very few had seriously contemplated the idea that something on the scale of a pocket watch, as opposed to a large clock, would ever be viable. After all, everyone knew what poor timekeepers watches were; when even the best could only manage about a minute a day, watches weren't even close contenders for longitude rewards.

That is not to say there wasn't interest in improving the performance of watches; there is evidence that, from the late 1740s, serious experiments were being made. In 1752, one noted London watchmaker, John Ellicott (1702/3–72), announced to the Royal Society that, following ideas he developed in 1748 (using bimetallic technology undoubtedly inspired by Harrison's well-known work), he had had a watch fitted with temperature compensation. However, Ellicott stated that before describing it to the Society he would wait to see whether it did indeed improve the watch's timekeeping. Nothing more was heard of this experiment and it is reasonable to conclude that the watch did not, after all, perform well. This conclusion is confirmed by a remark made by Harrison a few years later, in 1763, after his watch H4 was, to everyone's disbelief, going exceptionally well. He complained: 'but they still say a watch is [...] but a watch, and that Mr Ellicott has tried what a watch will do and that the performance of mine (though nearly to truth itself) must be altogether a deception'.

It is known that from at least the early 1750s Harrison was also experimenting with improvements in watches, although whether he began after hearing of Ellicott's work, or vice versa, we will probably never discover; for one who was, at the time, looking for alternative technologies to H3, Harrison may well have been prompted to look again at watches anyway.

The Jefferys watch. Completed in 1753, this watch was made to Harrison's design by John Jefferys. It is the watch shown in Harrison's hand in the portrait by Thomas King (pp. 32–3). The discolouration is due to fire damage sustained in the 1940s.

THE JEFFERYS WATCH

What is certain is that at about the same time, in 1751 or 1752, Harrison commissioned a London watchmaker, John Jefferys, to make him a watch following Harrison's own designs. The watch, which Harrison would have finished and adjusted himself, was evidently meant for experimental use, it might have been useful for his astronomical observing and clock testing, but was undoubtedly intended as a potential prototype marine timekeeper, depending on how well it performed. The watch was given a bimetal for temperature compensation, of course, but it had another feature that was much more significant for its timekeeping properties.

The movement of the Jefferys watch. With its high-energy balance it performed very well and was the breakthrough Harrison needed.

Up until the 1750s, watchmakers had always made the balance in their watches relatively light and small, so they oscillated with relatively low energy. The reason for this was that it was an absolute article of faith among professional watchmakers that, if a watch had run down and stopped, when it was wound up again, it must immediately spring into life – it must be 'self-starting'. One can imagine them saying, 'What's the use of a watch if you have to shake it to get it started?' The only way they knew to achieve this was by making the balance small and light, so that the escapement of the watch could easily push the balance to set the mechanism in motion.

Harrison's instincts told him, however, that in order for a watch to keep good time, its balance must oscillate with as much stored energy as possible, so that physical disturbances are proportionally smaller and less affecting. But to achieve this meant going completely against that cardinal rule in the watchmaker's book. The balance he designed for the Jefferys watch was relatively heavy (he gave it a rim of gold), somewhat larger than usual and beat faster than normal watches (the balance beats five times per second: it has, as watchmakers would say, 'an 18,000 train' – it beats 18,000 times an hour). The watch had a type of verge escapement – the standard watch escapement of the day – but one that was completely redesigned, enabling the balance to run at a much larger amplitude (with bigger swings than normal). All of this meant that the watch was by no means self-starting and, if the watch ran down, the case had to be given a little twist after winding up for it to start running.

With this redesign, Harrison had expected a moderate improvement in timekeeping, but to his amazement the difference was very great indeed and he began to realise the way forward. After all this time, he had in fact been working on a false assumption. The huge, slow-swinging oscillators of his pendulum clocks are fine if they are fixed to a very solid foundation, but if one wishes to start moving a timekeeper around, one has to rethink the technology altogether, as the large slow-beating balances are inherently unstable and prone to disturbance. The solution was, after all, in those 'hopeless' little things called watches, but it was vital that the oscillator within them was of the 'high-energy' type for them to succeed. This apparently simple discovery is one of Harrison's great achievements but, because it was an alteration to an existing part – the proportions of the balance and spring – as opposed to the introduction of a new device (like, for example, temperature compensation, or a different kind of escapement), this vitally important advance has been largely overlooked by horological historians.

The minutes of the Board for the meeting held on 18 June 1755 introduce the first hint to the outside world of what Harrison knew to be his breakthrough. As well as asking for money to continue H3, he requests support:

> … *to make two watches, one of such a size as may be worn in the pocket & the other bigger […] having good reason to think from the performance of one already executed according to his direction* [the Jefferys watch] *[…] that such small machines may be rendered capable of being of great service with respect to the Longitude at sea …*

Support was duly provided by the Board and Harrison began work immediately. Over the next four years he produced the one that was 'bigger', a watch that would in more recent times be known as 'H4', arguably the most important watch ever constructed.

THE 'LESSER WATCH'

Even today we are not sure whether the smaller watch, the 'one of such a size as may be worn in the pocket', was ever made. The Board certainly awarded Harrison money to make it, and we know he made a construction drawing of the movement, as this survives in the Clockmakers' Company collection.

Harrison also refers to it in his writings, calling it his 'lesser watch' (meaning smaller watch), which suggests it might have been made. So where is it? Nobody knows. The lesser watch has in recent years become legendary after its fictional 'discovery' in the final episode of the cult BBC sitcom *Only Fools and Horses*, broadcast at Christmas in 1996.

The main character, 'Del Boy' Trotter, 'sold' the watch through Sotheby's for £6,200,000, making him and his sidekick, Rodney, millionaires. Appropriately enough, the watch was 'purchased' on behalf of the National Maritime Museum, Greenwich, and all were happy ever after! Well, we can always dream.

That last episode was watched by 24.6 million viewers, more than the Queen's Christmas speech and the UEFA Euro 1996 Final that year! It gave the staff at Greenwich a few headaches afterwards, however: we were flooded with claims to have found the real lesser watch, something which, alas, has not yet come to pass.

Harrison's construction drawing for the movement of the 'lesser watch', c.1755. It is not known if the watch was ever completed.

HARRISON'S FOURTH TIMEKEEPER, 'H4'

Not surprisingly, then, in most respects the large silver-cased watch that Harrison made next – just 13 cm in diameter and weighing 1.45 kg – is completely different from the earlier marine timekeepers, though the same logical approach to design and the same construction methods are evident.

Both externally and to some degree internally, H4 looks like a very large contemporary pocket watch, even to the extent that it has pair cases (with an inner case for the movement and a protective outer case around it). Technically, though, it is different from an ordinary eighteenth-century watch in a number of significant ways. Apart from being exceptionally finely constructed, the movement's balance is much larger and of higher frequency. Like the Jefferys watch, H4's balance oscillates five times per second and, having the same escapement, the oscillations are much larger than in an ordinary watch. Temperature change was compensated for in the same way, by using a smaller version of H3's bimetallic strip. Unlike the Jefferys watch, H4 contains a miniature H3-type remontoire, rewinding eight times a minute, to ensure a constant drive to the escapement.

Harrison's fourth marine timekeeper, H4, made between 1755 and 1759, and considered by many to be the most important watch ever constructed. It is the foundation stone of all subsequent precision watches and chronometers.

60
5
10
15
20
25
30
35
40
45
50
55

The movement of H4. The beautifully engraved back of the movement covers the 'high-tech' balance and compensation. The signature records the additional contribution of Harrison's son William.

Although Harrison was unable to miniaturise his anti-friction devices, and H4 (like the Jefferys watch) required oil on all its bearing surfaces, jewelled bearings were fitted in many places to reduce friction to a minimum. The use of jewelled bearings was not Harrison's invention – they had been in use by London watchmakers since the beginning of the century – but the extent to which he used them was unprecedented, as was his use of diamond in the specially shaped pallets of the escapement. Harrison records that during H4's construction he had great difficulty forming the smooth curves he needed on these pallets (which were partly his means of ensuring that the balance and spring were isochronous), but he did succeed and, even today, in spite of close study, we are uncertain how.

The solid silver pair case is interesting. The inner, with the incuse (stamped) mark 'I H' and London hallmarks for 1758/59, might just have been made by Harrison himself. The inner case would have been needed first, to protect the movement during the process of final construction and adjustments, and, most importantly, to prevent prying eyes from seeing what was inside. The outer case, hallmarked for London 1759/60, is marked 'H T' and was possibly made by Henry Thompson, a 'small worker' and joiner in Silver Street, close to Harrison's house in Red Lion Square.

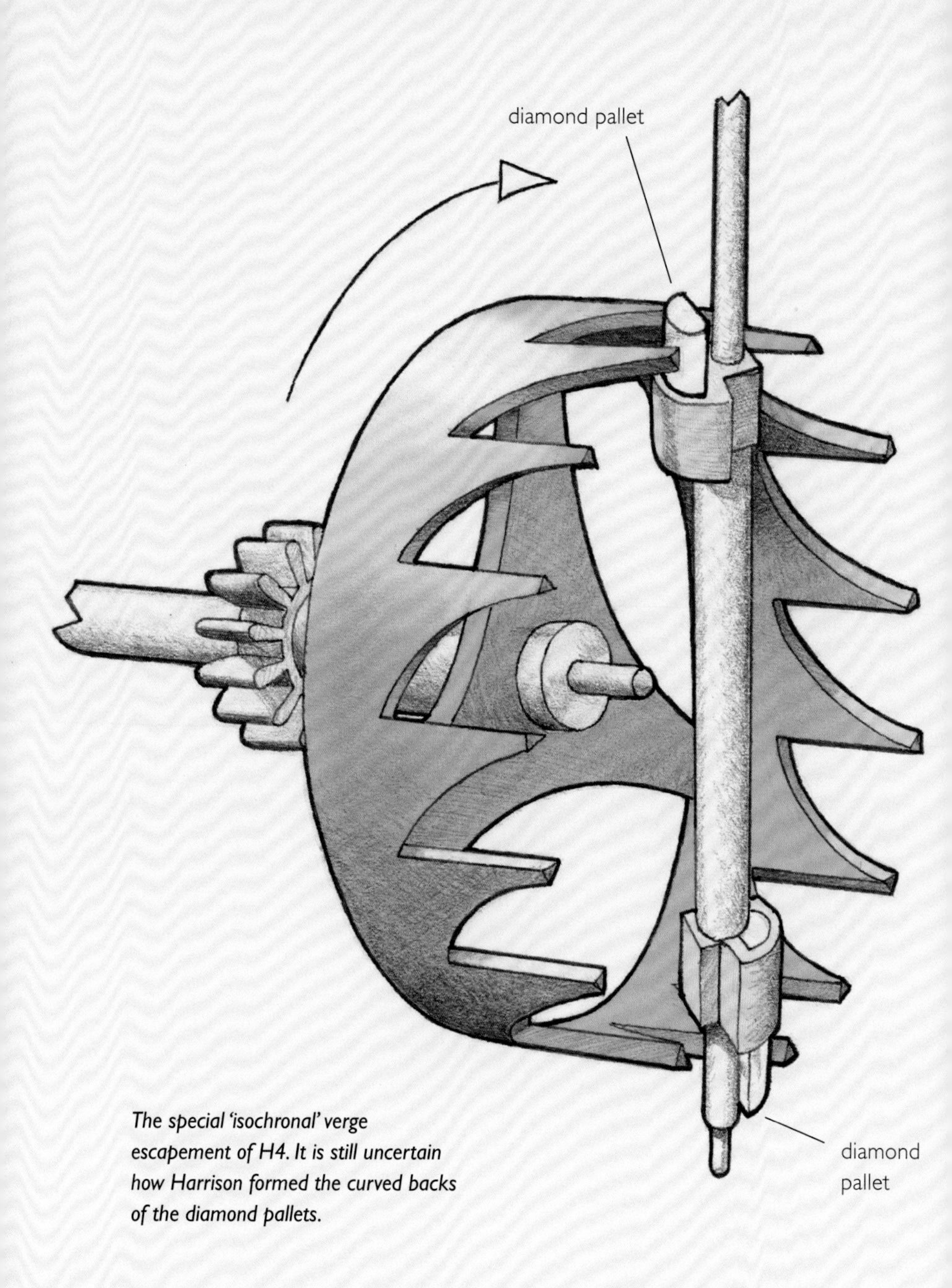

The special 'isochronal' verge escapement of H4. It is still uncertain how Harrison formed the curved backs of the diamond pallets.

TRIALS AND REWARDS

Harrison had made great progress with H4, but he was now not the only one coming closer to solving the longitude problem. At the very next meeting of the Board of Longitude, in March 1756, the Commissioners were told of some lunar tables recently completed by Professor Tobias Mayer of Göttingen University in Germany, which had been produced in parallel with the work of the Royal Observatory. Furthermore, the invention of the reflecting quadrant (the octant) by John Hadley (1682–1744) in 1731 (simultaneously with Thomas Godfrey (1704–49) in Philadelphia) made measuring the altitude of stars at sea easier and provided the sort of results that could be used in conjunction with the new tables. The Astronomer Royal himself, James Bradley (1693–1762), recommended that the Board reconsider the potential of the lunar-distance method for solving the longitude problem, as it was now sufficiently accurate and was becoming a truly viable alternative to timekeepers.

At the Board's meeting on 18 July 1760, Harrison asked for a trial of H3, hoping that it would be possible to send 'the watch' (H4) as well, as it had far exceeded his expectations. He said he needed one more winter to test its temperature compensation and this was allowed. In March 1761, £250 was allocated to John's son, William Harrison (born in 1728 and now in partnership with his father) so he could prepare for a trial voyage to Jamaica in charge of the timekeepers. After innumerable delays, on 18 November 1761, the *Deptford* sailed for the West Indies with William and H4 on board (in the event, H3 was not sent with them).

The watch had been set to exact local time at Portsmouth, by careful observation of the Sun at noon (a process known as 'equal altitudes'). As chance would have it, the crew's supply of beer (fresh water did not stay drinkable on long voyages) had become tainted en route, and the ship and its convoy urgently needed to put in at Madeira for provisions. Using H4, William was able to predict correctly that, at the time, they were nearly 100 miles closer to land than the navigating officer had judged, an achievement happily noted by the whole crew. Indeed, the Captain, Dudley Digges, was so impressed that he asked to purchase the next example of such a timekeeper Harrison might make.

The voyage on to Jamaica was equally successful for H4. They arrived at Jamaica on 19 January 1762 (the official end of the trial). After ascertaining exact local time at Port Royal by equal altitudes, it was evident that the watch had gone extremely well, though exactly how well would only be determined after careful calculations on returning home.

The passage back to England could not have been more different from the placid, happy voyage out. With few ships available at the time, William had been obliged to accept a cabin on the small sloop-of-war *Merlin* and the weather throughout the voyage was tempestuous, to say the least. At times, William had to cradle H4 in blankets to protect it from buffeting and seawater. He insisted, however, on keeping it going lest it should be thought too fragile to withstand use at sea. On arrival at Portsmouth its error, according to William's reckoning, was only 1 minute 54.5 seconds after a total period of 147 days. And calculations on the watch's performance during the official trial then showed it to have been in error by just 5.1 seconds during the whole voyage out! Needless to say, the Harrisons were very pleased.

It was a remarkable achievement but, alas, it was spoilt by one crucial oversight. There had been a failure to officially discuss and agree the *rate* of the watch. Even very accurate, reliable timekeepers do not usually keep exact time. It is extremely difficult to adjust a clock or watch so it does not gain or lose anything. As long as the amount is regular and predictable this really does not matter. If, for example, it gains 5 seconds each day (its 'rate') then one simply makes that much allowance every day and correct time can be deduced. However, for trials, this concept was entirely new. Since the Harrisons had not officially declared H4's rate before leaving port, there was absolutely no way for Board members to know if it was, indeed, as accurate as William claimed it was. The trial was therefore next to useless.

Predictably, the Commissioners of the Board of Longitude were not convinced by William's figures. At their meeting in June 1762, they announced that they were dissatisfied with the trial on a number of counts but, not surprisingly, top of the agenda was the question of applying a rate. In fairness to the Harrisons, it is clear that the whole trial had been badly organised from the outset and the Board was equally culpable in the uncertain results. It did, however, agree to award £2,500, of which £1,000 was to be paid only after another trial, terms to which the Harrisons reluctantly agreed.

From this point on, the attitude of the Board began to harden towards the Harrisons. The reasons for this are complex and not all the motives are entirely clear even today, but one element which certainly did not help was that, as the years went by, Harrison's old influential friends were dying (Graham in 1751 and Bradley in 1762, for instance) and he was losing his establishment support. Another problem was the definite prejudice of members of the Board against timekeepers as a solution; many years later one of the Commissioners claimed that the Board had never expected a timekeeper to qualify for an award at all. Even as late as 1792, Nevil Maskelyne (1732–1811), Astronomer Royal and member of the Board, was still asserting that lunar distances were more trustworthy for fixing longitudes, and it was only in the 1820s that the Royal Navy finally appeared to accept the chronometer as the best means for longitude determination.

For many Commissioners, the seeds of doubt must have been sown during the previous 19 years during which, in retrospect, Harrison surely appeared to have been floundering with H3. Now, as the contender suddenly seemed to be a watch, and looking much like an ordinary one, they were even less inclined to believe it could be successful. As quoted earlier, in 1763 Harrison himself complained bitterly about people claiming watches were just watches and could never succeed.

Considerations of character also should not be underestimated. We know, from his writings, that Harrison had some difficulty expressing himself clearly. From one point of view, this is hardly surprising. Many of the concepts and devices he was trying to describe had no name and he was constantly having to invent a new technical language. This, taken in combination with his relatively humble origins, his lack of formal education and his probable lack of social graces must have stood Harrison in poor stead before such an elite as the members of the Board of Longitude, especially as very large sums of money were at stake. As they saw it, the members of the Board had a huge responsibility: they were answerable to Parliament and the nation, and posterity would judge them harshly if they disposed of such a large sum of money without absolute proof that the winning method would be practical and genuinely solve the problem.

It should also be remembered that the make-up of the Board was constantly changing. Hardly any two meetings of the Commissioners were composed of the same people and, as the years went by, those less supportive of Harrison (and timekeepers) became a majority.

Add to this the emergence of the increasingly perfected lunar tables, a mathematical solution that undoubtedly appealed more to them as theoreticians, and perhaps the diminishing favour accorded the Harrisons is not so difficult to understand. Unfortunately, the hardening attitude of the Board was matched by mounting paranoia and suspicion on the Harrisons' side, evidenced by their increasingly argumentative and sullen attitude before the Commissioners.

At the meeting of the Board in August 1762, Harrison confirmed his agreement to a second trial of H4 to the West Indies. Remembering the dangers met during the return from the earlier trial, the Board also persuaded Harrison of the benefits of a formal disclosure of the technical make-up of H4, to which he tentatively agreed.

There must have been a certain breakdown in communication between the Board and Harrison on this matter, however, as it seems the Board's intention was always that details of H4 should eventually be made widely known for the benefit of the international scientific community, whereas the Harrisons evidently understood that what they were passing on was absolutely secret and, first and foremost, for British production.

FRENCH INTEREST

However, the Board of Longitude, believing a disclosure was now imminent, informed representatives from France that a visit to inspect the timekeeper was on the cards. In April 1763 the celebrated French astronomer Jérôme Lalande (1732–1807) visited Harrison and was able to get a look at the external appearance of the watch, but failed to ascertain any details of its interior. Then, in company with the mathematician Charles-Étienne Camus (1699–1768) and clockmaker Ferdinand Berthoud (1727–1807) he duly visited Harrison again in May, but this time the three were only shown the large timekeepers, Harrison stubbornly refusing to let them see H4 at all.

As a way of keeping the pressure on the Board, Harrison petitioned Parliament for a formal clarification of his situation. This 'clarification' took the form of an Act of Parliament, which came into force on 31 March 1763. It guaranteed that no one else could win the reward with a timekeeper until his had been properly tried. The Board, however, then augmented this by applying a set of rather severe requirements for the 'disclosure' of the construction of H4, adding that Harrison should oversee the making of two more watches. These would be tested to ensure that the information had been passed on effectively and to prove that others could satisfactorily make copies of H4.

As the making of other watches had never been even hinted at in the original Act of Parliament, which was still fully in force, Harrison refused to accept these extra conditions and there was an impasse. Neither party was willing to agree to a compromise and so the question of the disclosure was shelved while plans were made for H4's second trial to the West Indies. When the Board met on 4 August 1763 to make the necessary arrangements, the first item on the agenda was the dreaded question of deciding H4's rate. It was ultimately agreed that Harrison should be allowed to provide his own statement of what the rate should be. The astronomers were then nominated to carry out the necessary observations during the voyage, one being Nevil Maskelyne, an ardent supporter of the lunar-distance method of finding longitude and, not surprisingly, soon to be Harrison's bête noire.

In August 1763, Maskelyne was sent out to Barbados to set up the observation station from which the watch, on its arrival, could be compared to local time by equal altitudes. Maskelyne was also charged with making other land-based astronomical observations, in order to fix the longitude of the island with greater certainty. On his voyage out, Maskelyne had taken the opportunity to continue his tests of the lunar-distance method of determining longitude and was extremely proud of the accurate results he obtained.

Meanwhile, after making his declaration of H4's rate (gaining 1 second a day in ambient temperature) to the Admiralty on 24 March 1764, William Harrison and Thomas Wyatt, a companion, departed with H4 in the ship *Tartar* from Spithead at Portsmouth on 28 March. As with the earlier trial, William predicted the arrival at Madeira with extraordinary accuracy. Captain Sir John Lindsay presented William with a certificate to that effect and all seemed to be going well, until William's arrival at Barbados.

No sooner had William stepped ashore, than rumours reached his ears that Maskelyne had been boasting about the success of the lunar-distance method he had employed to find longitude on his passage out from England. Understandably, William felt that Maskelyne was not the man to compile an unbiased report on the timekeeper and, after heated discussion, it was decided to appoint one of the other astronomers, Charles Green (1734–71), to share Maskelyne's work.

A Sixth Rate on the Stocks, *by John Cleveley the Elder, 1758. The painting shows (right) a sixth-rate ship of the line, virtually identical to the* Tartar, *and (in the foreground) a sloop-of-war identical to* Merlin, *both ships used for the trials of H4.*

In the event, the trial was another astonishing success story for H4. The average computation put the watch's error at just 39.2 seconds after the voyage of 47 days.

This was three times better than the performance needed to win the full £20,000 longitude reward. Whatever may have been said and done before, the Board should now have acknowledged that the terms of the original Act of Parliament had been met and the reward had been won. Sadly, the Commissioners saw matters

differently and were not yet ready to pay anything. First, they raised the old spectre of the disclosure and stated that they would pay half the total, £10,000, once Harrison had made a proper exposé of H4's mechanism, on oath, to a specially appointed committee. The details would then be published for the benefit of the world at large. Second, the Board implied that the watch was a fluke and that others of the same kind should be made and tested. Third, the Board then had these requirements sanctioned in a further

Act of Parliament, which also included the demand that all four timekeepers should be handed over once the £10,000 had been paid to Harrison.

At this stage the Harrisons' relations with the Board were at an all-time low. To fuel their paranoia, the man they were learning to distrust above all others, Nevil Maskelyne, had been appointed Astronomer Royal in 1765 and was therefore an ex-officio member of the Board. For several weeks, Harrison refused to negotiate any point of the Board's proposals, but the mood of the Board was equally militant. Realising that they would get nowhere if he did not compromise, John Harrison finally agreed to sign the oath and disclose the inner workings of the watch.

Anticipating this, the Board had already arranged a panel of six experts to inspect the instrument and witness the explanation. The panel consisted of three well-respected practical watchmakers – Thomas Mudge (*c.*1715–94), William Matthews (dates unknown) and Larcum Kendall (1719–90) – the Reverends William Ludlam (*c.*1717–88) and John Michell (1724–93), and the London instrument maker John Bird (1709–76). Kendall had been apprenticed to John Jefferys and, as Jefferys had died in 1754, it is possible he had helped Harrison with the making of some parts of H4.

The disclosure meeting could hardly have been a relaxed affair under these circumstances, but – to add to the tension – the Board appointed Maskelyne to oversee the presentation. Beginning on Wednesday 14 August 1765, the panel gathered at Harrison's house in Red Lion Square and watched him dismantle H4. Members of the panel heard Harrison's answers to the technical questions put to him and he also provided drawings of the watch for further explanation. On 22 August, the panel was apparently satisfied and each member signed a certificate stating that the disclosure had been complete (though later at least one of them expressed doubts as to having learned everything they needed).

The Board met again on 28 October 1765 and granted Harrison enough money to make up £10,000, the first half of the full reward. In return, it insisted on the other stipulation: that the timekeepers should all be handed over. After numerous pleas from Harrison the Commissioners grudgingly allowed him to hold on to H1, H2 and H3. They insisted on having H4, though, as they intended to ask Larcum Kendall to produce a copy of it, to reassure them that workable copies could be made, entirely above and beyond the requirements stipulated in the 1714 Act of Parliament.

At last Harrison had half the money. But, for him, it was only the whole reward and, most importantly, the recognition that he had produced a practical solution to finding the longitude, that mattered. Since the failed visit in 1763, the French clockmaker Ferdinand Berthoud had, via a third party, remained in contact with Harrison, and now, apparently feeling cheated of his rightful rewards, Harrison agreed to negotiate with the French for details of H4. He received Berthoud in London in early 1766 but, expecting £4,000 for his disclosure, he sent him away again on discovering that Berthoud had come with only £500! Not to be put off, Berthoud then approached one of the Board's own disclosure panel, the watchmaker Thomas Mudge.

Unfortunately for Harrison, Mudge, a man of great integrity, was happy to oblige with information, believing it to be the Board's intention that the knowledge should be disseminated. Unsurprisingly, the Harrisons were not happy on discovering what they considered to be a breach of trust and published a broadsheet on the matter. To Harrison's good fortune, Berthoud did not make good use of what Mudge had conveyed and the machines he later made in Paris bear only surface similarities to Harrison's work.

THE

PRINCIPLES

OF

MR. HARRISON'S TIME-KEEPER,

WITH

PLATES OF THE SAME.

PUBLISHED BY ORDER OF

THE COMMISSIONERS OF LONGITUDE.

LONDON:

PRINTED BY W. RICHARDSON AND S. CLARK;

AND SOLD BY

JOHN NOURSE, AND MESS. MOUNT AND PAGE.

M. DCC. LXVII.

The Principles of Mr Harrison's Time-Keeper, *1767. Although limited in the information it contained, this book was highly influential in the fundamentals it revealed to the next generation of chronometer makers.*

The whole matter was to some extent made irrelevant a year or so later, as in April 1767 the Board published *The Principles of Mr Harrison's Time-Keeper*. It was translated into French within a few weeks. Although at the time the publication was said to have been of insufficient use to those wishing to continue developing marine timekeepers, the evidence today shows that it was, in fact, immensely influential and arguably the most seminal of all publications in the history of the chronometer.

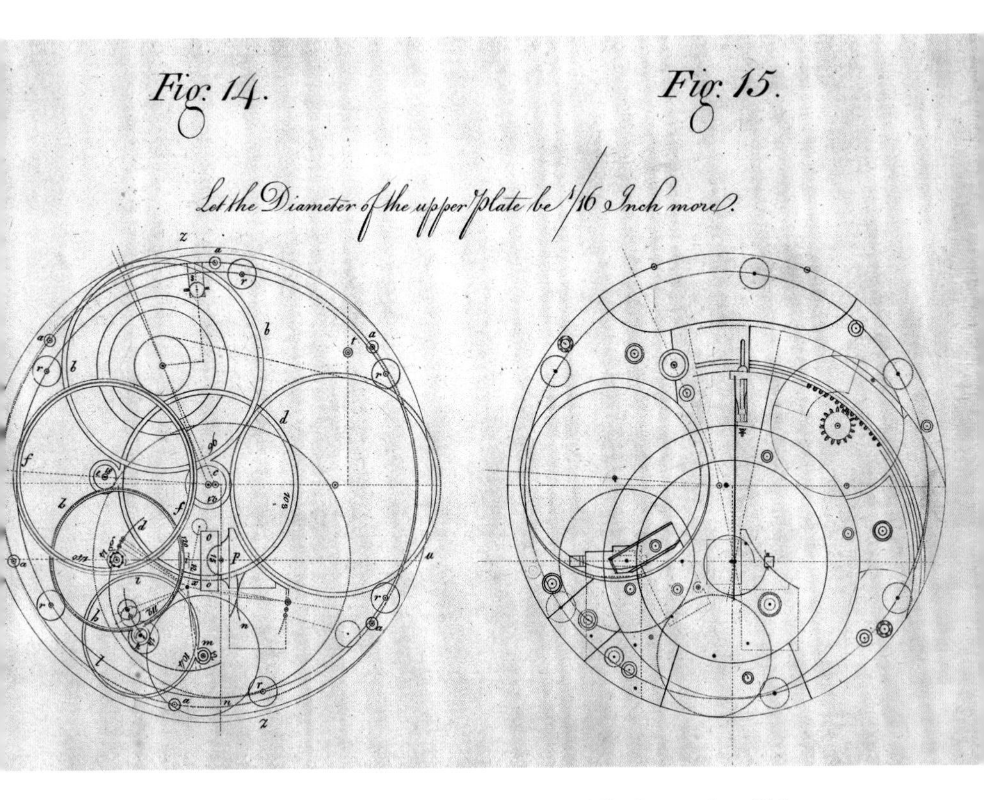

Two construction drawings for the movement of H4 that appeared in The Principles of Mr Harrison's Time-Keeper, *1767.*

THE SECOND £10,000

Harrison now knew that, in order to satisfy the Board of Longitude, he had to produce at least two other watches similar to H4 and have them tested. He wrote to the Board suggesting two alternatives: he could make the two himself, for which he would need a grant of £800, or he could set up a factory and employ other workmen to make any number of them under his supervision. The latter proposal was contingent on his being paid the £10,000 in advance.

Unfortunately, Harrison was in no position to bargain with the Board. In its reply of 26 April 1766, it flatly refused either proposition, telling him to make the two timekeepers (particularly difficult since it had taken H4 from him). A decision would be made about the reward only after the two timekeepers had been tested. Evidently the Board had decided to move the goal posts on the original 1714 Act of Parliament and now considered that the full reward for one successful trial of one timekeeper was not sufficient. The Commissioners were sending a clear message: in receiving half the reward, they considered Harrison had done well and should not be awarded any more unless he absolutely met the requirements of the new Act as well. Understandably, Harrison was increasingly unhappy with this new direction.

Knowing that he would not give up the challenge of gaining the whole longitude reward, they decided to test all four timekeepers at Greenwich, a completely pointless exercise as Harrison's claim lay in the performance of H4, not the larger machines. Nevertheless, on 25 May 1766, without any advance warning, Nevil Maskelyne, of all people, turned up at Red Lion Square to collect H1, H2

Portrait of the fifth Astronomer Royal, Nevil Maskelyne, attributed to John Russell, c.1776. *Drawn in black and red chalk on blue paper.*

and H3 from Harrison. To add insult to injury, Maskelyne arrived with an un-sprung cart – the sort of transport that could do more damage to Harrison's timekeepers than years at sea. Harrison was extremely reluctant to advise how they should be best moved, lest he be implicated if they were damaged in transit. Predictably, one of the timekeepers (H1) was indeed damaged.

H4 GOES ON TRIAL AT THE OBSERVATORY

In the meantime, H4 began its ten-month trial at the Observatory; a trial which, it seems, was destined to go badly from the very outset. Owing to the fact that H4 had not been cleaned or properly readjusted since its dismantling for the disclosure, and because it had been left in storage for months at the Admiralty before going to Greenwich, the timekeeper was wholly unprepared for a fair trial.

To make matters worse, the watch did not have entirely appropriate treatment while under Maskelyne's care at the Observatory, being subjected to extremes of temperature and positions of running far beyond what it was designed for. Predictably, it did not perform at all well and the conclusions of Maskelyne's published report are not surprising: '… Mr Harrison's watch cannot be depended upon to keep the Longitude within a degree in a West India Voyage of Six Weeks …' The fact that Harrison's watch had been dependable on at least one such voyage was completely obscured by the so-called formal trial. Harrison was incensed, immediately publishing a virulent response, *Remarks on a pamphlet lately published…*, but to no avail.

View of the Royal Observatory, Greenwich, from the South East, *artist unknown, c.1770. H4 was tested here by Nevil Maskelyne in 1766.*

HARRISON'S FIFTH TIMEKEEPER, 'H5'

Greatly embittered but still determined to see the project to its end, Harrison and his son began to make the first 'copy timekeeper' to qualify for the remaining reward money. On 11 April 1767, Harrison approached the Board, again requesting the loan of H4 so that he could make an accurate copy. It refused, saying that Larcum Kendall needed H4 in order to make his copy. The Board members also informed Harrison, to his horror, that they had devised a whole new, ten-month plan for testing the watches.

In spite of this further setback, Harrison (now in his mid-70s) and his son continued working for over two years on the fifth timekeeper, known today as 'H5'. Meanwhile, Kendall's good progress on his copy had the adverse effect of persuading the Board that all Harrison's difficulties to date had been merely procrastinations. Kendall's watch, now referred to as 'K1', was completed in 1769 and was inspected in early 1770 by the same panel that had seen H4. William Harrison was also present at the demonstration and agreed that the copy was exceptional. Another request to the Board that both K1 and H5 be considered as the 'two timekeepers' for an official trial was also turned down. The Harrisons were told, in no uncertain terms, that it was they who must make both timekeepers, a demand that yet again bore no relation to the original Act of 1714, which remained in force. No wonder the Harrisons were unhappy.

By 1772, further finishing and adjusting of H5 had provided only one further timekeeper. The idea that John Harrison himself, at the

Harrison's fifth timekeeper, H5, completed in 1770. Although much plainer in appearance, the watch was technically very similar to H4, with slightly improved temperature compensation.

age of 79, might sit down and make a second watch was clearly absurd. Father and son had reached the end of the road and desperate measures were needed if further progress was to be made. The Board was obviously no longer prepared to dedicate much more time to the Harrisons; its interests lay elsewhere. With Maskelyne's *Nautical Almanac* now being published every year and the navigational instrument, the octant, newly available in a much-improved form, the sextant, the lunar-distance method had become entirely viable.

It is thus true to say that the members of the Board of Longitude were prejudiced against timekeepers as a workable means of finding longitude. It is not, however, true that members of the Board engaged in a vendetta against Harrison, though Maskelyne's actions during the trial were wholly unacceptable. The Board was now rejecting the terms of the original Act of Parliament, but was perhaps simply doing what it believed to be right; in spite of the evidence of the trials, it just didn't believe timekeepers would ever be as reliable a solution as the lunar-distance method.

Besides, even if Harrison's timekeepers did prove themselves useful, they needed to be much less expensive before the Admiralty could afford to equip the fleet. K1 cost the Board £500 – the equivalent of about £25,000 in today's money; Harrison's prototype certainly appeared to be a very expensive design and in those days it was not understood that in technological development, larger-scale production would see costs reduce, in spite of Harrison's assurance that they would.

KING GEORGE III

Sensing he had almost run out of options, as a last resort John decided to appeal to the highest authority in the land, the King himself. Harrison and his son William had had an audience with George III in 1764 and they knew he was following their fortunes with interest. An approach was made to the King on 31 January 1772 by letter, via his private astronomer at Richmond, Stephen Demainbray (1710–82). William Harrison requested an opportunity for H5 to be put on trial by the King himself at his private observatory. William was summoned for interview at Windsor and asked to expand on some of the details. At this interview the King is said to have remarked: 'these people have been cruelly treated'. He then apparently exclaimed: 'By God, Harrison, I will see you righted!'

The King was keen to help and agreed to put H5 on trial at Richmond from May to July of that year as a form of independent test. After a false start, apparently caused by leaving the watch too close to some magnetic lodestones, H5 performed superbly. Its daily rate of variation over the whole 10 weeks averaged out at less than a third of a second per day. Both the King and Demainbray were very impressed and the Harrisons believed their own personal trial was nearing its end.

Hoping that the involvement of the King might cause some change of heart in the Commissioners, Harrison approached the Board of Longitude again on 28 November, citing H5's good behaviour and asking for the remaining £10,000. It was a vain hope. The Board replied that only an official trial would suffice and that 'no regard will be shewn to the result of any trial made of them in any other way'.

Portrait of King George III, by Sir William Beechey, c.1800. The King was a great supporter of the Harrisons at the time that members of the Board of Longitude were objecting to the awards of money to them.

Probably at the suggestion of the King himself, Harrison now formally approached the Prime Minister, Lord North, with the full story. This appeal had the interesting effect of causing the Speaker of the House of Commons to instruct the Board of Longitude to reassess the case in William Harrison's presence, witnessed by two MPs who were Harrison supporters. At this meeting, the Commissioners posed a number of specific, formal questions to William, who, somewhat churlishly, gave only abrupt answers, mostly in the negative. When finally asked why he refused now to submit just one timekeeper to a trial, he replied: 'For the following reasons ... Loss of time, Expense attending it, Uncertainty of reward afterwards and I think I can employ my time better.' The Board's minutes continue: 'He then withdrew.' With that, all contact between the Board and the Harrisons ended.

Harrison's petition to Parliament required further redrafting before it could be guaranteed a successful result. Finally, the recommendations of a specially appointed parliamentary Finance Committee were accepted by the House on 21 June 1773. The Act of Parliament entitled 13 George III, chapter 77, duly received Royal Assent and Harrison was awarded £8,750. This was not the £10,000 for which Harrison was hoping, but if one adds it to all the sums he had from the Board before, including expenses, the final sum actually totals over £23,000. Harrison had, at last, received the majority of the great longitude award.

George III's private observatory at Richmond. H5 was tested here between May and July 1772 and performed exceptionally well.

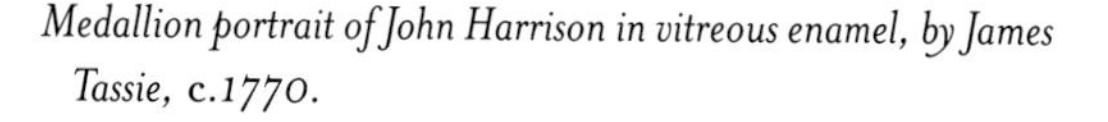

Medallion portrait of John Harrison in vitreous enamel, by James Tassie, c.1770.

The interesting question remains as to whether Harrison had actually 'won' the full reward money. The members of the Board of Longitude were opposed to the final payment, and it could be argued that without their approval one cannot say the full reward had been made. It might also be said, as no doubt Harrison did, that all the sums, excepting the first £10,000 and Parliament's £8,750, were not reward but were covering his costs; but it is not clear from the Board's deliberations what were, and were not, considered expenses. However, it was agreed by Act of Parliament that Harrison should receive the money he did, and, whether the Commissioners liked it or not, that money was all paid to him through the Board of Longitude's own accounts, so it can be reasoned that the majority of the reward was effectively received.

Perhaps more important to Harrison was that the reward be seen to be his. It still rankled with him that the members of the Board had not approved the granting of the money, but here was some recognition that John Harrison had solved the longitude problem.

Less than three years later, on 24 March 1776, Harrison died at Red Lion Square. It was his 83rd birthday. He was buried in the churchyard of St John's, Hampstead, north of the city, and his large stone monument, under which his son William is also buried, survives there to this day.

CONCLUSION

In 1772, Larcum Kendall's copy of H4, K1, was given the most exacting trial imaginable when it was issued to Captain James Cook (1728–79) on his second voyage to the South Seas. It performed magnificently. From his own log of the voyage, we can read of Cook's steady conversion towards belief in the timekeeper. From being a traditionalist and one who had learned to trust the lunar-distance method on his first voyage, the second voyage saw Cook gradually won over by using an instrument in which he had faith.

References to 'our trusty friend the watch' and 'our never failing guide' speak volumes from a man of Cook's abilities and experience. K1 was used by Cook on both his second and third voyages to chart many Pacific islands and north-eastern Pacific coastlines, amply demonstrating the reliability of both concept and the hardware. It is not known whether Harrison heard of K1's success after James Cook's second voyage, but it is highly likely he would, as an old man, have been informed soon after Cook returned in July 1775. One certainly hopes so.

LUNARS

In fact, once reliable tables were available in the late 1760s, the lunar-distance method of finding longitude also proved entirely workable and both systems had advantages. Lunars enabled longitude to be found, whereas if a chronometer stopped it was useless until

Captain James Cook, by Nathaniel Dance, 1776. It was Cook's highly successful trials of K1 in his second and third voyages to the South Seas which proved to the world that Harrison's design was sound.

a known longitude could be found to set it again. On the other hand, chronometers enabled longitudes to be determined at times of the month when lunars were not possible. In practice, navigators employed both lunars and chronometers for finding longitude during the rest of the eighteenth century and into the nineteenth, though by the 1820s chronometers were established as the principal means of navigation for most of the world's navies.

THE MARINE CHRONOMETER

The story of the marine chronometer, as such instruments became known, does not end here. Following Harrison's proof that such a timekeeper was possible, and with some of H4's essential design features published and available to watchmakers, a number of London makers found ways to simplify Harrison's design while preserving the fundamentals that ensured its good performance. It was not entirely in London that these design simplifications took place.The evidence suggests that one important concept, the 'free' or 'detached' escapement, central to modern chronometer design, was first thought of by the celebrated French pioneering clockmaker Pierre Le Roy (1717–85).

Head and shoulders above the rest, among this next generation of makers, however, was John Arnold (1736–99). Encouraged by Nevil Maskelyne, who presented him with a copy of Harrison's *Principles* immediately after publication, Arnold was responsible for the majority of the design improvements in the modern marine chronometer. It then only remained for watchmaker Thomas Earnshaw (1749–1829) to standardise the form, which enabled the concept to be made quickly, cheaply and in large numbers.

Thus, England remained at the forefront of chronometer and precision-watch production up to the end of the nineteenth century. It would not be an exaggeration to say that without Harrison's pioneering work, Britain's foreign trade would not have developed so extensively and its Empire could not have expanded as rapidly as it did. For nearly two centuries Britannia did indeed 'rule the waves', and Harrison, in no small measure, enabled it to do so.

Above left: *Watchmaker John Arnold, by Mason Chamberlin, c.1765. It was Arnold alone who contributed the most, in detail, to the developed chronometer after Harrison.*

Above right: *Watchmaker Thomas Earnshaw, by Sir Martin Archer Shee, c.1808. Earnshaw's principal contribution was to standardise the form of the modern marine chronometer.*

HARRISON'S LEGACY

Harrison's timekeepers themselves suffered much less august fates. Ever since the disastrous trials under Maskelyne, the timekeepers remained at Greenwich. Save one impromptu cleaning session in the 1830s, they were virtually untouched until the 1920s. They lay in store, dirty, dismantled and decaying. It was only when Lieutenant-Commander Rupert T. Gould (1890–1948), a polymath and a keen amateur horologist, expressed an interest in them that they once again saw the light of day.

Gould described their condition: 'All were dirty, defective and corroded, while No.1, in particular, looked as though it had gone down with the *Royal George* and had been on the bottom ever since.' He continued:'I could not bear to see them in this condition. It seemed to me such a futile, tragic ending to a great adventure. They were the first accurate marine timekeepers ever made – the life work of an original genius who was also an Englishman – and here they were, discarded […] forgotten […] buried. Surely they deserved a better fate.'

Gould wrote the definitive book on marine chronometers, a work profoundly and thoroughly researched and one that, only in the twenty-first century, with so much new historical study taking place, is now beginning to show its age. Despite having no formal horological qualifications, Gould was allowed to undertake the restoration of the timekeepers during the 1920s and 30s. Initially he was only permitted to clean the corrosion off H1, which was in by far the worst condition, but, after successfully restoring its appearance (in 1920), Gould was allowed to restore H4 (1921),

H2 (1923–25), the R.A.S. regulator (1927–29), H3 (1929–31) and then finally to complete a full restoration of H1 (1931–33).

A brilliant and remarkable polymath, Gould was fanatical in his interest in Harrison and was one of the twentieth century's finest antiquarian horologists. So determined was he to see Harrison's legacy restored to its former glory, it not only took him much of the rest of his life, but also caused him to lose nearly everything he held dear. This magnificent obsession of his ultimately contributed to the total breakdown of his marriage and the loss of his home, custody of his children, his closest friendships and even his job. He paid a very high price and it should be remembered that it is largely thanks to Gould's heroic efforts to save Harrison and his timekeepers from neglect that we are able to enjoy seeing the timekeepers in their

Lieutenant-Commander Rupert T. Gould, photographed at his home in Epsom in 1924. He sits next to H3 and has one of the balances of H2 on his lap.

splendid, working condition to this day. It must also be said that, after Gould's death, the care of the timekeepers was undertaken by the Chronometer Section of the Ministry of Defence, which continued Gould's good work and improved upon one or two of his less elegant repairs.

So well was all this work carried out over previous years that today the large timekeepers run constantly, without fear of significant deterioration, and almost without breakdown. All that is required of their current custodians, the horological staff of Royal Museums Greenwich, is their daily winding, though there is still much to learn about their design and construction, and an ever-watchful eye is kept on their conservation to ensure they are preserved, with minimal change, for centuries to come.

Harrison's timekeepers on display at the Royal Observatory, Greenwich.

HARRISON COMMEMORATED

In recent years Harrison's fame has also undergone a renaissance. In 1993 the tercentenary of his birth was celebrated by the total refurbishment of the Royal Observatory's historic galleries, and the first edition of this book was produced on that occasion. The same year, Harrison medals were struck by the Royal Mint, the British Post Office released a celebratory set of Harrison stamps and there was an important conference held at Harvard University on the subject of longitude. The conference resulted in the publication of

The memorial to John Harrison in Westminster Abbey, made by Joanna Migdal and unveiled on 24 March 2006 by Prince Philip. Harrison's name is engraved across a bimetal, running due north–south across the tablet.

the fine book *The Quest for Longitude*, but also inspired Dava Sobel to write her bestseller *Longitude,* which brought Harrison's name to a much wider audience. In addition to the comedy show *Only Fools and Horses* in 1996 (see p. 84), Harrison was also commemorated at the beginning of this millennium in a number of ways.

The tercentenary saw the start of petitions to have Harrison's great contribution recognised with a memorial in Westminster Abbey, finally bearing fruit in 2006 with a memorial tablet unveiled in the nave. Appropriately, this was right next to the grave of his old friend George Graham, who was buried in the same tomb as Thomas Tompion. With this memorial, and his own wonderful monuments in the form of the timekeepers preserved at Greenwich, we may rest assured that the name of John Harrison will be remembered and revered for many hundreds of years yet. Finally, in 2017, the whole collection of marine chronometers at Greenwich, including all of the Harrison timekeepers, were studied and catalogued for posterity.

SUGGESTED READING

R. T. Gould, *The Marine Chronometer*, Potter, London, 1923. Although now 100 years old, this is still an important book on marine chronometers.

H. Quill, *John Harrison: The Man Who Found Longitude*, Baker, London, 1966. Currently the only full biography of Harrison, and very well researched.

D. Sobel, *Longitude (new edition)*, Walker, New York, and HarperCollins, London, 2005. The bestseller that tells the Harrison story in an exciting and accessible style.

W. Andrewes (ed.), *The Quest for Longitude*, Harvard University, Cambridge, MA, 1996. The proceedings of the Harvard University Longitude conference of 1993, containing a series of erudite and definitive chapters on the history of longitude.

J. Betts, *Time Restored*, Oxford University Press/National Maritime Museum, 2006. The biography of R. T. Gould and the story of the Harrison timekeepers and their 20th-century restorations.

J. Betts, *Marine Chronometers at Greenwich*, Oxford University Press/National Maritime Museum, 2017. All the Harrison timekeepers at Greenwich are fully described and analysed in this comprehensive book.

PICTURE CREDITS

Images (except where stated below) © National Maritime Museum, Greenwich, London. Additional images are reproduced by kind permission of the following:

Front cover (bottom) © iStock

4, 29, 73, 75 Jonathan Betts

20–1 National Maritime Museum, Greenwich, London, Palmer Collection. Acquired with the assistance of H.M. Treasury, the Caird Fund, the Art Fund, the Pilgrim Trust and the Society for Nautical Research Macpherson Fund

22 Graphic by Matt Windsor

30 National Maritime Museum, Greenwich, London. Acquired with the assistance of the Heritage Lottery Fund, The Art Fund, the Worshipful Company of Clockmakers and the Garfield Weston Foundation

31 (both) National Maritime Museum, Greenwich, London. Acquired with the assistance of the Heritage Lottery Fund, the NACF, the Worshipful Company of Clockmakers and the Garfield Weston Foundation

32–3 Science Museums Group

37, 51, 109 Image: © National Maritime Museum, Greenwich, London / Courtesy of the Clockmakers' Company Museum

38, 44, 80, 81, 85 The Worshipful Company of Clockmakers' Collection, UK © Worshipful Company of Clockmakers /Bridgeman Images

41 Dr John C Taylor OBE (also courtesy of the Earl of Yarborough)

43 Andrew King

77 Image: © National Maritime Museum, Greenwich, London / On loan from the Royal Astronomical Society

78 Image: © National Maritime Museum, Greenwich, London / Collection of Don Saff, USA. By permission of Don Saff

98–9 National Maritime Museum, Greenwich, London, Caird Collection

105 National Maritime Museum, Greenwich, London, presented by the descendants of Nevil Maskelyne

112, 118 (right) National Maritime Museum, Greenwich, London. Caird Fund

113 R J F Brothers

116 National Maritime Museum, Greenwich, London, Greenwich Hospital Collection

118 (left) ©The Trustees of the British Museum. All rights reserved

120 Reproduced with the permission of Sarah Allan

122 © 2022 Dean and Chapter of Westminster

INDEX

ABOUT THE AUTHOR

JONATHAN BETTS MBE, FSA, FBHI, FIIC, FRSA was appointed Senior Horology Conservator at the National Maritime Museum in 1979, Senior Curator of Horology in 2000 and Curator Emeritus in 2015. He is Vice Chairman of the Antiquarian Horological Society and a Past Master of the Worshipful Company of Clockmakers (2014). In 1989 he received the NMM Callender Award, in 2002 the Clockmakers' Company's Harrison Gold Medal, in 2008 the BHI Barrett Silver Medal, in 2012 the BQ Chinese 'Watch Culture' award in Beijing, in 2013 the Plowden Medal (RWHA) for his contribution to horology conservation, and in 2015, the Swiss Prix Gaia for his achievements in historical research. He received an MBE for services to horology in 2012.